STUDENT WORK

Prepared by
John E. Neely

Lane Community College (ret.)
Eugene, Oregon

MACHINE TOOL PRACTICES

FOURTH EDITION

Richard R. Kibbe

Oxnard Community College
Oxnard, California

■

John E. Neely

Lane Community College (ret.)
Eugene, Oregon

■

Roland O. Meyer

Lane Community College
Eugene, Oregon

■

Warren T. White

San Jose State University
San Jose, California

■

PRENTICE HALL, Englewood Cliffs, New Jersey 07632

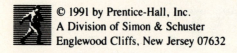

© 1991 by Prentice-Hall, Inc.
A Division of Simon & Schuster
Englewood Cliffs, New Jersey 07632

All rights reserved. No part of this book may be
reproduced, in any form or by any means,
without permission in writing from the publisher.

Printed in the United States of America

10 9 8 7 6 5 4

ISBN 0-13-544065-3

Prentice-Hall International (UK) Limited, *London*
Prentice-Hall of Australia Pty. Limited, *Sydney*
Prentice-Hall Canada Inc., *Toronto*
Prentice-Hall Hispanoamericana, S.A., *Mexico*
Prentice-Hall of India Private Limited, *New Delhi*
Prentice-Hall of Japan, Inc., *Tokyo*
Simon & Schuster Asia Pte. Ltd., *Singapore*
Editora Prentice-Hall do Brasil, Ltda., *Rio de Janeiro*

CONTENTS

INTRODUCTION 1

PROJECTS
    Drill and Hole Gage 3
    Turning and Threading Project 9
    Making a Center Punch 17
    C-Clamp 25
    Paper Punch 33
    Parallel Clamp 43
    Tap Handle 51
    Die Stock 59
    Center and Taper Sleeve 67
    Internal Thread 73
    Multiple Lead Threads 77
    Acme Threads 83
    Carbide Tool Exercises 87
    Hydraulic Jack 93
    Vee Blocks 129
    Precision Vise 143
    Spur Gear 161
    Cylindrical Grinding 165
    Tool and Cutter Grinding 171
    Final Term Project 177

ALTERNATE PROJECTS
    Pin Punch Set 181
    Wheel Puller 187
    Machine Vise 199

TABLES
    Decimal and Metric Equivalents of Fractions
        of an Inch 211
    Tap Drill Sizes for Unified and American Standard
        Series Screw Threads 212
    Metric-Inch Conversion Table 214
    Inch-Metric Measures 214
    Useful Formulas for Finding Areas and Dimensions
        of Geometric Figures 215
    Allowances for Fits of Bores in Inches 215
    Tapers and Corresponding Angles 216

INTRODUCTION

This workbook is designed to be used in conjunction with the textbook, <u>Machine Tool Practices</u>. As you encounter each phase of a project or exercise, you will be instructed on the worksheet to study and complete Post-Tests for certain units in the textbook. If you are directed to refer to certain units in the textbook, this means you should use the units merely to refresh your memory for the operation, since you will have already completed the study and Post-Tests for those units in an earlier project. It is absolutely necessary that you follow the routine and sequence of study. Without the related theory, it would be very difficult, if not impossible, and definitely hazardous to you to try to complete the project in this workbook. Do not make the mistake of putting off doing your "paperwork" until after the project is finished, since that would defeat the purpose of the learning system and you would miss some important training. In most cases, credits and grades will not be given on your project until the required Post-Tests are completed and turned in to your instructor. Without your instructor's approval, you should not begin the next project or exercise.

Your textbook contains answers to the Self-Tests in each unit in the appendix. The Self-Tests will help you to check your knowledge of the subject matter before taking the Post-Test. Also, in the textbook appendix, are a number of helpful tables and useful formulas. For your convenience, some useful information and tables have also been included in the appendix of the workbook, since it will always be nearby or at the machine on which you are working.

The project instruction sheet organizes your study in the textbook and in the workbook worksheets. When a unit of study in the textbook has been completed and the Post-Test taken, the worksheet for that unit that is in your workbook should also be completed, unless your teacher has given you instructions to omit that worksheet. When you have completed your project, take it to your instructor for evaluation. Also tear out the project grading sheet which precedes the worksheet and give it to your instructor.

These projects and exercises were planned to give you some training and experience in using machine tools in various ways to produce useful parts and mechanisms. In machine shop work, tolerances are very important. On the beginning projects the tolerances are very broad; standard tolerances (plus or minus 1/64 inch) are used. However, in later projects the tolerances become closer; in some cases, within tenths of one thousandths of an inch.

It is important for you to know how to follow drawings without error to make precision parts, since this is what you must do as a machinist. You will be able to achieve this ability without great difficulty if you follow the procedures given to you in the textbook and this workbook.

Since the main objective of the major lathe project is to give you more experience in using the lathe in a variety of operations, the project must be more complex than any you have completed up to that point. This project will also bring you briefly into some basic milling operations. You must therefore plan to spend a great deal of time completing this project, perhaps as much as 100 to 200 clock hours in the shop.

Two major lathe projects are provided in this workbook so that you may have a choice. Your instructor may also have other selections for you to choose from. In any case, your selection must be made with the permission of your instructor and with his or her advice.

For any number of reasons, you may not wish to take the time to do the most time-consuming or challenging project and so may wish to choose the less difficult one. These two projects are placed in the order of difficulty to produce: first, the hydraulic jack project, the most difficult; and second, the wheel puller, the least difficult. The wheel puller is treated as an alternate project and may be found in the alternate project section.

The precision vise project also has an alternate, the machine vise. This, too, gives a choice between a more exacting project requiring extremely close tolerances and precision grinding techniques and one whose tolerances are broader and which requires only milling machine finishes. Again, the choice must be subject to the advice and supervision of your instructor. Your instructor may or may not assign greater or lesser credits or grades to these projects; you should find out before choosing.

The final term project is a mechanism of your own design, planned and drawn to scale. The choice of project will be subject to your instructor's acceptance and permission while it is still in the sketching stage. It must then be drawn to scale with assembly and detail drawings. The device must then be made in the machine shop, according to the drawings. Some examples of these final projects are drill presses, power hacksaws, hydraulic gear pumps, and wood lathes.

It is my hope that the textbook, Machine Tool Practices, and this workbook will help give you a better understanding of the machining trade and make you a more knowledgeable and capable machinist on the job.

J.E.N.

MACHINE TOOL PRACTICES

Name _____ Date _____

PROJECT 1. DRILL AND HOLE GAGE

Project Evaluation (To be filled out by the instructor):

	Grade	
	Letter	Percent
1. Follows drawing (dimensions and tolerances)	_____	_____
2. Machining finishes	_____	_____
3. Mechanism or tool operates satisfactorily	_____	_____
4. General workmanship	_____	_____
Total grade	_____	_____

Comments:

Signed: _____
(Instructor)

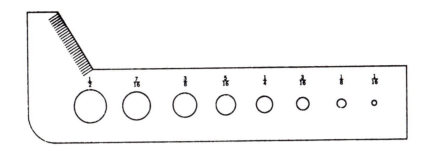

PROJECT 1. DRILL AND HOLE GAGE

Objectives

1. Learn general safety in the shop.
2. Learn how to use power cutoff tools and hand cutting tools in bench work.
3. Learn to measure and lay out workpieces accurately.
4. Prepare for machining operations.
5. Be able to select drills correctly from any of several series.
6. Be able to offhand-sharpen twist drills on a pedestal grinder.
7. Be able to drill holes accurately with a drill press.

Outline for Study

Prior to starting each procedure for this project, study and complete the units in the textbook and the Post Tests for:

1. Safety instruction: Section A, Unit 1.
2. General shop information: Section A, Units 2 and 3.
3. Using the steel rule: Section C, Unit 2.
4. Preparing to use machine tools: Section F, Units 1, 2, and 3.
5. Cutoff sawing: Section G, Units 1 and 2.
6. Layout: Section E, Unit 1.
7. Using hacksaws and files in bench work: Section B, Units 3 and 4.
8. Selecting drills and drilling procedures: Section H, Units 1 through 3.

Procedures

Begin the procedures for this project by completing the following worksheets.

Worksheet 1. Cutoff sawing.
Worksheet 2. Layout.
Worksheet 3. Making the drill gage.
     A. Sawing the shape of the drill gage.
     B. Filing the surfaces of the drill gage.
     C. Drilling the holes.

For drill grinding procedure and drill selection, refer to Section H, Unit 3.

WORKSHEET 1. Cutoff Sawing

Exercises

1. Obtain some scrap material and make some practice cuts using the reciprocating power hacksaw and horizontal band cutoff machines.

2. Cut material for the drill gage project using the reciprocating or band cutoff machine.

WORKSHEET 2. Layout

Exercise

Obtain a piece of scrap material. Properly prepare the material and do the following layout. Alternatively, do only the layout of the drill gage as shown in the textbook, Section E, Unit 1.

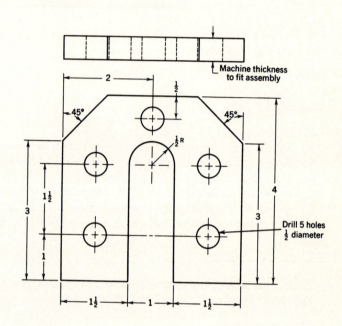

WORKSHEET 3. Making the Drill Gage

A. Sawing the Shape of the Drill Gage.

Materials

Hacksaw $\frac{1}{8}$ x 2 inch cold finish (CF) steel bar
Vise

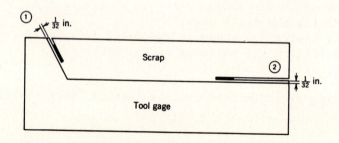

Procedure

1. The length of the material should be cut off $\frac{1}{16}$ inch longer than the finish size of the drill gage.
2. Make the angular saw cut 1, leaving $\frac{1}{32}$ inch of stock for filing as shown in the drawing. Hold the work so that the saw cut is in a vertical position. Do not saw to the end of the line. Leave $\frac{1}{16}$ inch.
3. Clamp the material so that saw cut 2 is vertical. Keep the cut as close to the vise jaws as possible to prevent vibration. Saw as far as possible with the saw blade in the normal position. When the limit of the frame has been reached, turn the blade 90 degrees and continue sawing.
4. Leave $\frac{1}{32}$ inch outside the layout line for filing. If the kerf begins to wander in or out and you cannot correct it by angling the saw, cut off the scrap piece from the outside edge to the end of the saw cut and start a new cut.
5. Continue the cut until it joins the angular cut. Be careful and do not undercut at the corner.

B. Filing the Surfaces of the Drill Gage

Materials

Selection of files   Vise
Previously cut drill gage as shown

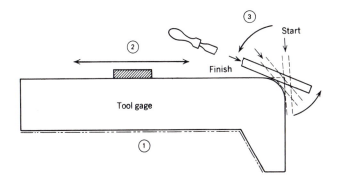

Procedure

1. Rough filing on the sawed surface, side 1, may be done by cross filing. If a large amount of stock needs to be removed, use a double cut file. Leave $\frac{1}{64}$ inch for finish filing.
2. Using a single cut file, all the straight surfaces may be finished by draw filing, side 2. Check frequently with a rule or straight edge to make sure that the surface is flat. The high spots are easily seen and should be filed until almost no light is seen between the rule and the part.
3. Rough file the radius, side 3, with a double cut file leaving $\frac{1}{32}$ inch for finishing.
4. With a single cut file, begin as shown as side 3. As you file with a forward stroke, move the handle downward.
5. Check frequently with a $\frac{1}{2}$ inch radius gage. Continue to file on high spots until almost no light is seen between the gage and the part.

C. Drilling the Hole

<u>Materials</u>

No. 2 center drill plus the following drills: $\frac{1}{2}$, $\frac{7}{16}$, $\frac{3}{8}$, $\frac{5}{16}$, $\frac{1}{4}$, $\frac{3}{16}$, $\frac{1}{8}$, $\frac{1}{16}$ inch
Countersink   Drill press vise
Parallels   Cutting oil

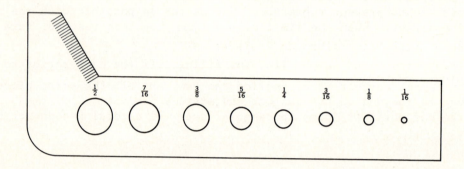

<u>Procedure</u>

1. Test all of the drills you are going to use on similar material to your drill gage. Resharpen if needed.
2. After laying out and center punching for the hole location, as in the illustration above, set up the drill gage in the vise with parallels arranged under each edge so that the largest drill can clear when you drill through.
3. Place the center drill in the chuck and tighten. Spot it with the machine off in the punch mark for the $\frac{1}{2}$ in. hole. Clamp the vise without moving it.
4. Set the correct speed and drill into the workpiece so there is a slight chamfer. <u>Do not</u> spot drill for the $\frac{1}{16}$ in. hole, but spot drill all the other holes. You might enlarge the $\frac{1}{16}$ in. hole oversize.
5. Change to the $\frac{1}{16}$ in. drill and drill that hole.
6. Change to the $\frac{1}{8}$ in. drill and drill the $\frac{1}{8}$ in. hole and pilot drill all the others.
7. Continue drilling all the remaining holes, taking care to change to the correct speeds for each drill size.
8. Using the countersink, lightly chamfer both sides of each hole.
9. The reference marks used for checking lip lengths may now be put on with a scriber and rule, a specially prepared stamp, or a scriber and division setup on a milling machine. Check with your instructor for the method you should use.
10. Turn in the grading sheet and the finished project for your instructor's evaluation.

MACHINE TOOL PRACTICES

Name _____ Date _____

PROJECT 2. TURNING AND THREADING PROJECT

Project Evaluation (To be filled out by the instructor:

	Grade	
	Letter	Percent
1. Follows drawing (dimensions and tolerances)	_____	_____
2. Machining finishes	_____	_____
3. Mechanism or tool operates satisfactorily	_____	_____
4. General workmanship	_____	_____
Total grade	_____	_____

Comments:

Signed: _____
(Instructor)

MACHINE TOOL PRACTICES   11

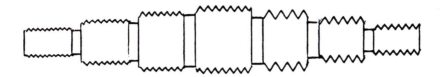

PROJECT 2. TURNING AND THREADING

Objectives

1. Learn to use precision measuring tools.
2. Learn to grind lathe tools for turning and threading.
3. Be able to use a small engine lathe for external turning, grooving, and threading.

Outline for Study

Prior to starting each procedure for this project, study and complete Post-Tests for:

1. Precision measurement: Section C, Units 1, 3, and 4.
2. Introduction to the lathe: Section I, Units 1, 2, 4, and 5.
3. Grinding a lathe tool: Section I, Unit 3.
4. Facing and centerdrilling: Section I, Unit 6.
5. Turning between centers, undercutting, and chuck work: Section I, Units 7 and 9.
6. Cutting threads: Section I, Units 10 and 11.
7. Measuring threads: Refer to Section I, Unit 11.

Procedures

Begin the procedures for this project by completing the following worksheets.

Worksheet 1. Grinding a lathe tool bit.
Worksheet 2. Centerdrilling and facing.
Worksheet 3. Measuring threads.

WORKSHEET 1. Grinding a Lathe Tool Bit

Grind one acceptable right-hand turning tool bit as shown in Section I, Unit 3.

WORKSHEET 2. Centerdrilling and Facing

Material

Number 3 centerdrill   Piece of $1\frac{1}{4}$ in. diameter mild steel HR round $7\frac{1}{8}$ in. long.
Right-hand turning tool

Procedure for Centerdrilling

1. Set up the $1\frac{1}{4}$ in. diameter workpiece in a chuck on a lathe and face one end without leaving a center stub.
2. Set up a Number 3 centerdrill in the drill chuck on the tailstock and set the correct speed. Using cutting oil, drill the center hole to the desired depth. About three-quarters of the way into the countersink section would be a good depth in this case.
3. Break the corner with a file; that is, deburr the sharp outer edge left by machining.
4. Turn the part around in the chuck and repeat steps 1, 2, and 3.

Procedure for Turning between Centers

The specifications and tolerances as given in the drawing below must be observed and held unless your instructor has a different set of allowances.

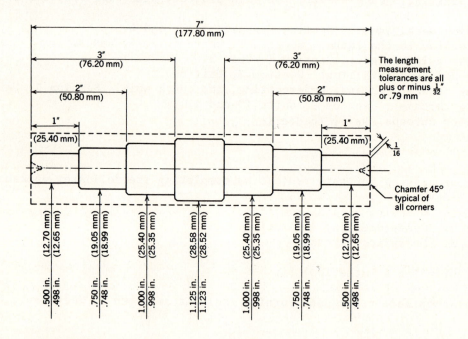

1. Set up the lathe for turning between centers. Make sure the centers and sockets are clean and the centers are well seated.
2. Apply high pressure lubricant to the center hole at the tailstock end. Ordinary oil will not work as it will run out and the center will heat up and be damaged.
3. Slip on the dog with the bent tail pointed toward the drive plate.
4. Place the workpiece between the centers.
5. Put the dog in place and tighten the set screw. Make sure the bent tail is clear of the bottom of the slot.

6. Adjust the dead center so there is no end play, yet with sufficient clearance so the bent tail of the dog is free to move when the part is rotated. Tighten the tailstock binding lever. <u>Note</u>: Further adjustment of the dead center will be necessary when the work heats up from machining.
7. Fasten a right-hand turning tool in a straight or left-hand toolholder or in any appropriate holder provided on your lathe. Adjust the holder so the tool will not gouge the work if it should slip from heavy cutting. Set the tool for height.
8. Set the lathe to the correct RPM for the work.
9. Adjust the feed for roughing; 0.008 to 0.012 in. would be appropriate for a small lathe.
10. <u>Always check to see that the workpiece or dog is clear by turning the spindle by hand before using the power</u>. <u>Wear safety glasses</u>. Turn on the machine.
11. The first roughing cuts should extend past the center section of the workpiece so that a distance of roughly $4\frac{1}{2}$ in. should be turned from the tailstock end.
12. Touch the tool to the work, zero the crossfeed dial, and take a skim or clean-up cut of about 0.005 in. depth for about $\frac{1}{2}$ in. Turn off the machine. <u>Leave the dial setting as it is</u>.
13. Check this turned size with your micrometer.
14. In most turning operations, it is standard practice to leave all diameters of a turned part slightly oversize to allow for any warping before taking the final finish cut. This operation is called "roughing." About 0.030 in. is sufficient to leave for finish. Most small lathes should have sufficient power to make the first diameter with one roughing cut and one or two finishing cuts. For example, let us assume the $1\frac{1}{4}$ in. diameter is now 1.245 in. and the finish diameter will be 1.125 in. Thus, 0.120 in. is the total amount of stock to be removed from the diameter of the middle section. About 0.030 in. should be left for finishing, leaving 0.090 in. for the roughing cut.
15. In this case, you would now set your depth of cut for 0.045 or 0.090 in. diameter, depending on the type of graduation on the cross feed collar of your lathe. Complete the roughing cut for a length of about $4\frac{1}{2}$ in. Check for taper and, if it is more than 0.001 in., reset the tailstock. (See Section I, Unit 8, "Alignment of Lathe Centers.")
16. The next size is a 1 in. diameter that has a shoulder. All diameters are roughed on both ends before taking any finish cuts.
17. Scribe or mark a line on the workpiece 3 in. from the tailstock end.
18. Make a roughing cut to within $\frac{1}{16}$ in. of the line. This allowance is left for later clean up of the shoulder. Increase lathe speed to compensate for smaller diameter.
19. Scribe a line 2 in. from the tailstock end and repeat step 18. Scribe a line 1 in. from the tailstock and repeat step 18.
20. Turn the workpiece end for end and rough turn each step as before.
21. Set up a right-hand finishing tool or a modified right-hand turning tool (by grinding a larger nose radius on it) in the toolholder.
22. Set the lathe for a finishing feed and speed.
23. Run the tool in until it touches the workpiece and move it to the starting position. Add 0.003 in. more and zero the micrometer collar. Take a light cut for about $\frac{1}{4}$ in.
24. Check the diameter with the micrometer caliper. It should now be between 0.020 and 0.030 in. oversize. Set the cross feed micrometer collar for the appropriate amount and take a trial cut $\frac{1}{8}$ in. long. Check for size with a micrometer and adjust if necessary, so that the finish size will be within the tolerances given on the drawing. Finish the cut.

25. When all of the diameters are finished on one end, turn the tool so that a square shoulder can be made. Pick up the cut (just touch the finish diameter) near the shoulder and hand feed to the line. Back out the tool with the cross feed handle to square the shoulder. Turn the tool SCEA about 45 degrees and chamfer the edges.
26. Turn the work end for end and repeat.

Procedure for Undercutting

Before the threading operation is begun, a means of clearing the threading tool should be arranged. Undercutting to the depth of the thread is one such means.

1. Using either a parting tool or a specially ground tool, make an undercut for each thread equal to its single depth plus 0.005 in. Use cutting oil. Make the undercuts $\frac{3}{16}$ in. wide. If the tool chatters, check for too much toolholder or tool overhang; slow down the machine.

2. The following formula will give you the single depth for undercutting Unified threads:
$$d = P \times 0.613 \quad \text{or} \quad d = \frac{0.613}{n}$$

where
   d = single depth
   P = pitch
   n = number of threads per inch (TPI)

Since there is no undercut for the $1\frac{1}{8}$-12 thread, consider the other six diameters and threads only.

Fill out the following list for the single depth of undercutting on each:

14 TPI _____   8 TPI _____

16 TPI _____   10 TPI _____

20 TPI _____   13 TPI _____

Use this list as a guide for depth of undercutting as you undercut each step on the project shown below.

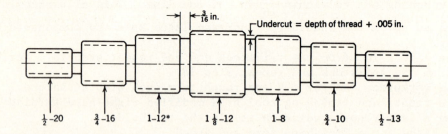

* <u>Note</u>: Use 1-12 UNF or 1-14 UNS (former American Standard threads). Many shops do not have 1-12 UNF nuts in stock, but have only 1-14 UNS nuts to use for testing the thread. Check with your instructor to see which thread pitch you should use.

3. Set up a parting tool and touch it to the finish diameter you wish to undercut. Zero the micrometer collar and back off the tool. Now move the tool to the shoulder and make the cut. You may need to double the calculated depth on some lathes that read directly for diameters. For single depth reading cross feed collars, use the figures as they are. You may have to take two cuts side-by-side to get the $\frac{3}{16}$ in. width.

Procedure for Threading

1. For this operation, you will need to grind the opposite end of your lathe tool bit to make a 60 degree threading tool as shown in Section H, Unit 3. Use a center gage to check it. Provide for a zero or neutral back rake on the tool, depending on the type of toolholder you are using.
2. Set the compound at 29 degrees to the right, off the cross slide index line.
3. Determine the depth the tool is fed into the work when the compound slide is set at 29 degrees for the following threads. The formula for Unified threads is

$$d = P \times 0.708 \quad \text{or} \quad d = \frac{0.708}{n}$$

Fill out the following table for the infeed when threading each step.

12 TPI _____   8 TPI _____

14 TPI _____   10 TPI _____

16 TPI _____   13 TPI _____

20 TPI _____

4. Set up the threading tool, making sure it is on center and square to the work axis. Use the center gage for aligning the tool to the work.
5. Select a slow RPM and set the quick-change gears to give you 12 threads per inch so you can do the $1\frac{1}{8}$ in. diameter thread first.
6. Put the feed change lever or its equivalent in neutral.
7. Set the compound micrometer collar on zero.
8. Bring the tool in contact with the work on the largest diameter using the cross slide. Then set its dial on zero.
9. Move the carriage to the right until the tool is clear of the work; feed in the compound 0.003 in. Apply cutting oil to the work. <u>Always</u> use cutting oil when threading.
10. When the index mark on the thread dial is in line with the proper position on the dial, engage the half-nut lever. Keep your hand close to it so you can disengage the lever as soon as it has cut the full length of the thread.
11. With the cross slide, back the tool away from the work one or two turns. Return the carriage by hand to the right, where you started.
12. Return the cross slide to its zero position.
13. Feed the compound another 0.005 in (single depth); 0.010 in. (double depth).
14. Repeat steps 10, 11, 12, and 13 until 0.010 in. remains of the infeed dimension. Then feed only 0.002 in. per pass. The last pass should be made without additional feed in order to clean up the thread. Check the thread for size. When you are nearing the last few passes, use a standard nut for testing the thread. It should fit snugly, but you should be able to turn it on with your fingers. When turning between centers, using a dog, the workpiece may be removed for testing without disturbing the tracking of the thread if the dog is returned to the same slot.

15. Repeat steps 5 to 14 <u>with the changes made for each specific thread</u>. When reversing the workpiece between centers, protect the thread with two nuts locked on each other, or use a threaded dog.

WORKSHEET 3. Measuring Threads

Material

Screw thread comparison micrometer and a precision plug gage, or
Three-wire set and micrometer, or
Screw thread micrometer
Finished threading project

Procedure

1. Measure all of the thread diameters and pitches.
2. Record the type of measuring system you are using and the measurement of each thread.
3. Record the correct measurement for each thread and note the difference.

Type of measuring instrument: _____

TPI	13	10	8	12	14	16	20
Your measurement							
Correct measurement							
Total error oversize							
Total error undersize							

4. Turn in the grading sheet and the finished project for your instructor's evaluation.

MACHINE TOOL PRACTICES

Name _____   Date _____

PROJECT 3. MAKING A CENTER PUNCH

Project Evaluation (To be filled out by the Instructor):

	Grade	
	Letter	Percent
1. Follows drawing (dimensions and tolerances)	_____	_____
2. Machining finishes	_____	_____
3. Mechanism or tool operates satisfactorily	_____	_____
4. General workmanship	_____	_____
Total grade	_____	_____

Comments:

Signed: _____

(Instructor)

MACHINE TOOL PRACTICES

PROJECT 3. MAKING A CENTER PUNCH

Objectives

    1. Learn to identify ferrous and nonferrous metals.
    2. Learn to turn a taper using the compound rest.
    3. Learn to make a knurl.
    4. Be able to harden and temper plain carbon tool steel correctly.

Outline for Study

    Prior to starting each procedure for this project, study and complete Post-Tests for:

    1. Identifying metals: Section D, Units 1 and 2.
    2. Machining the center punch: Section I, Unit 13.
    3. Hardening and tempering the punch: Section D, Unit 3.
    4. Noncutting hand tools: Section B, Unit 2.

Procedures

    Begin the procedures for this project by completing the following worksheets in this workbook.

    Worksheet 1. Identifying metals.
    Worksheet 2. Machining the center punch.
    Worksheet 3. Hardening the center punch.
    Worksheet 4. Tempering the center punch.

WORKSHEET 1. Identification of Various Metals

<u>Materials</u>

Box of numbered specimens and a similar set of known and marked specimens

<u>Procedure</u>

Correctly identify by comparison, using various tests described in the textbook.

Item Number	Test Used	Kind of Metal
1.	_____	_____
2.	_____	_____
3.	_____	_____
4.	_____	_____
5.	_____	_____
6.	_____	_____
7.	_____	_____
8.	_____	_____
9.	_____	_____
10.	_____	_____

WORKSHEET 2. Machining the Center Punch

<u>Material</u>

A lathe and tooling   A piece of $\frac{3}{8}$ in. diameter drill rod
Aluminum protective sleeve or $\frac{3}{4}$ in. diameter rod from which to make one

<u>Procedure</u>

(a) Make a center punch (Figure 1) turning all the tapers with the compound rest method.
(b) Make an aluminum protective sleeve (Figure 2) if one is not available. It is needed to protect the knurled surface from the chuck jaws when the taper is turned. The material for this is an aluminum $\frac{3}{4}$ in. diameter rod $1\frac{1}{2}$ in. long with a $\frac{3}{8}$ in. diameter hole drilled in the center and slotted with a hacksaw.

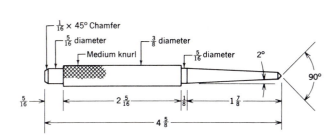
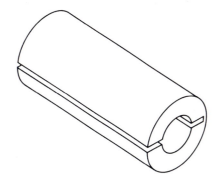

Figure 1.   Figure 2.

1. Cut off a piece of $\frac{3}{8}$ in. diameter drill rod, $\frac{3}{4}$ in. longer than the finished length of the center punch. (This allows for center drilling at the ends and removing the center holes after knurling.)
2. Use a three-jaw chuck to hold the drill rod; let it extend 1 in. from the chuck.
3. Fasten a number 2 or number 3 centerdrill in the tailstock drill chuck.
4. Centerdrill the workpiece; use cutting oil.
5. Exchange the drill chuck for a tailstock center.
6. Remove the work from the chuck and mark the length of the required knurl on it, then return the work into the chuck and let it extend only far enough out of the chuck so that you can knurl the marked off part.
7. Select a medium diamond knurling tool and fasten it in the tool post. Set it as close as possible to the chuck side of the compound rest with the least amount of overhang. See Section I, Unit 9, for knurling steps.
8. Move the carriage toward the chuck until the knurling tool just clears the chuck jaws. Turn the chuck by hand to check for clearance.
9. With the lathe turned on, set your depth of cut. (Try to get a finished knurl with the first or second pass; drill rod work hardens.)
10. Secure the knurled piece of drill rod in the protective collet (as shown in Figure 2) in the three-jaw chuck. Let $2\frac{1}{2}$ inches extend from it.
11. Select and set the correct RPM (the cutting speed for high carbon steel in the annealed condition should be about 70 SFM). Note that this RPM will be quite high, perhaps to top speed of the lathe. If you try to turn this slender rod at a low speed, it will most likely ride up on the tool and bend.
12. Fasten a right-hand tool in your tool post, so you can face off your workpiece and remove the existing centerhole. <u>Caution</u>: This is not a very rigid setup; work that is supported only on one end should not normally extend from the chuck more than four or five times its diameter.
13. Face the center punch so the distance from the end to where the knurl starts is 2 in. or less. Feed by hand using the cross slide; take very light cuts.
14. The 90 degree angle is machined at this point. Swivel the compound rest to 45 degrees. Tighten the carriage lock bolt. Take light cuts (0.020 to 0.030 in. deep). Hand feed by turning the compound rest handle until a sharp point is formed. Slow, even feed makes a smooth finish.
15. The $\frac{5}{16}$ in. diameter is now turned for a length of 2 in. in from the point to a shoulder.

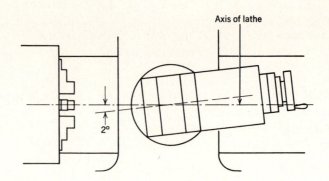

Figure 3.

16. Reset the compound so it will cut an angle of 2 degrees from the centerline of work (Figure 3).
17. Turn the compound rest handle counterclockwise until it is at the end of its travel.
18. Move the carriage so that the tool clears the point just turned by $\frac{1}{16}$ to $\frac{1}{8}$ in. and lock the carriage to the ways.
19. Take light cuts and turn a taper that measures $1\frac{7}{8}$ in. long from the point. Make a slow, even feed for the last cut to obtain a finish.
20. Remove the workpiece from the chuck and mark the other end for the length of $\frac{5}{16}$ in. diameter and the knurl length.
21. Rechuck the center punch so that the mark is visible outside the collet and turn the $\frac{5}{16}$ in. diameter to the mark.
22. Face to the required length.
23. Chamfer the edge. This completes the machining of your center punch.

Note. The center punch must be hardened and tempered before it can be used. See the next two worksheets and follow the procedures there for heat-treating your punch. Show the completed center punch to your instructor before heat-treating it.

WORKSHEET 3. Hardening the Center Punch

Materials

    Electric or gas furnace   Quenching media (oil or water)
    Tongs   Safety equipment (face shield and gloves)

Procedure

1. Since the center punch is a small tool, it should be placed in a furnace that has already been brought up to the correct temperature.
2. Determine the best place to grip the part with the tongs so that it will not be damaged when it is red hot. Use the proper shaped tongs.
3. Make sure you can grasp the part in the furnace to remove it in the proper orientation to enable you to quench it straight in.
4. Heat the end of the tongs so they will not remove heat from the punch.
5. When the punch has become the same color as the furnace bricks, remove it and immediately quench it <u>completely under</u> in the bath.
6. Agitate it up and down or in a figure-8 motion.

7. Be sure it has cooled below 200°F (93°C) before you remove it from the quench.

<u>Note</u>. If no furnace is available, a torch may be used with a color temperature chart.

Conclusion

1. Did the punch get hard? Test with an old file in an inconspicuous place. If a hardness tester is available, a quick test could be taken. (For information on hardness testing, see the unit on Rockwell and Brinell hardness testers.)
2. Is the part as hard as it should be? If not, check with your instructor.

WORKSHEET 4. Tempering the Center Punch

Material

| Fine emery cloth | Electric or gas furnace or oxy-acetylene torch |
| Tongs | Safety equipment |

Procedure for Tempering in the Furnace

1. Polish all smooth surfaces of the part with emery cloth and remove all oil. A cold furnace should be brought up to the correct temperature.
2. The center punch is then placed in the furnace for about 15 minutes.
3. Remove and cool in air.
4. When this method is used, the striking end of punches, chisels, and other striking tools should be given additional heating with a torch until they are a blue color on that end. This is done to ensure the safety of the user by preventing the struck end from shattering, since it is softer when tempered to blue.

Procedure for Tempering with a Torch or Hot Plate

1. When using a torch or hot plate, make sure that heat is applied to the body of a punch or part of the tool that can be softer. Allow the heat to travel slowly out to the cutting edge. In this way the colors may be observed.
2. See that the striking end of a tool is blue or gray before the proper color arrives at the cutting end.
3. When the proper color has arrived, quickly cool the piece in water. <u>Do not</u> delay or it will be overtempered.

Conclusion

1. Are the colors right according to Table 2, "Temper Color Chart," Section D, Unit 3, in the textbook?
2. If possible, recheck hardness and compare with Table 1, Section D, Unit 3.
3. Leave the temper colors on your project so your instructor can evaluate it.
4. Turn in the grading sheet with your project.

MACHINE TOOL PRACTICES

Name _____   Date _____

PROJECT 4. C-CLAMP

<u>Project Evaluation</u> (To be filled out by the instructor):

                                                                             Grade

                                                        Letter   Percent

1. Follows drawing (dimensions and tolerances)   _____   _____

2. Machining finishes   _____   _____

3. Mechanism or tool operates satisfactorily   _____   _____

4. General workmanship   _____   _____

                                        Total grade   _____   _____

<u>Comments</u>:

                                                  Signed: _____
                                                                              (Instructor)

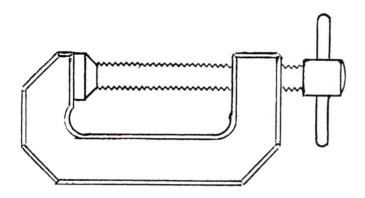

PROJECT 4. C-CLAMP

Objectives

1. Learn to use semiprecision layout methods.
2. Learn to use a vertical bandsaw correctly.
3. Be able to drill, tap, and cross drill on a drill press.
4. Be able to do turning and threading operations.

Outline for Study

Prior to starting each procedure for this project, study and complete Post-Tests for:

1. Layout procedures for the body: Section E, Unit 2.
2. Making the body: Section G, Units 3 and 4; Section H, Unit 4.
3. Making the screw: Section I, refer to Units 10 and 11.
4. Making the swivel: Section I, refer to Unit 9.

Procedures

Begin the procedures for this project by completing the following exercises and worksheets in this workbook.

Worksheet 1. Making the body.
   Exercise 1. Preparing a bandsaw blade.
   Exercise 2. Using the vertical bandsaw.
Worksheet 2. Making the screw.
Worksheet 3. Making the swivel.

WORKSHEET 1. Making the Body

Materials

$4\frac{1}{16}$ in. of $\frac{3}{4}$ x 2 in. CR flat bar $4\frac{1}{16}$ in. of $\frac{1}{2}$ in. dia. CR shaft

about 2 in. of $\frac{3}{4}$ in. CR shaft   2 in. of $\frac{3}{16}$ in. dia. CR rod

EXERCISE 1. Preparing a Bandsaw Blade

1. Obtain a worn out band blade and cut several suitable lengths. Practice end grinding, welding, annealing, and weld grinding procedures.
2. Practice folding band saw blades into small coils suitable for storage.

EXERCISE 2. Using the Vertical Bandsaw

1. Use the job selector and list all of the pertinent sawing data for the following materials:
    (a) Mild steel   (b) Aluminum   (c) Cast iron
2. Familiarize yourself with the band machine controls.
3. Obtain some scrap material. Consult the job selector and set proper band velocity. Make several straight and contouring cuts to familiarize yourself with band machine operation.
4. Use the band machine on your C-clamp after it is laid out.

Procedure for Making the Body

1. Obtain $\frac{3}{4}$ by 2 in. CR steel and cut off a $4\frac{1}{16}$ in. length in the power saw. Deburr.
2. Clean the material in solvent.
3. Using a vise with jaw caps, file both ends square and flat to 4 in. long. Note that all dimensions are given in Drawing II at the end of this project.
4. Apply layout dye to one side. Let it dry.
5. Follow the layout procedure as shown in Section E, Unit 2, pages 257 to 260 in your textbook.
6. Lay out the lines where you will saw. Deepen the punch marks with a center punch.
7. Drill the $\frac{1}{2}$ in. holes.
8. With the vertical band saw, cut along the layout line, leaving about $\frac{1}{32}$ in. of material for filing.
9. File to the marks, keeping the work square and flat. Carefully finish file to the point of tangency of the $\frac{1}{2}$ in. holes. Be careful not to nick the radius with the edge of the file.
10. File the large angles on the corners flat and square.
11. File a $\frac{1}{16}$ in. chamfer on all edges except for the anvil which is deburred only.
12. Lay out for the $\frac{3}{8}$-24 threaded hole.
13. Follow the procedures given in Section H, Unit 4, pages 378 to 379 in your textbook for drilling and tapping the $\frac{3}{8}$-24 hole.

Note: An alternate method to that given in the textbook of holding the tap in the drill press is to clamp the tap directly in the drill chuck and start it by hand by turning the chuck. Do not turn on the power. When you have completed several turns of the tap, release the tap from the drill chuck, leaving the tap still in the workpiece, and take the work to a bench vise, or leave it in the drill press vise, and complete the tapping by hand. This will insure a straight tapped hole. Clean up the body and threads.

WORKSHEET 2. Making the Screw

Material

CR shaft $\frac{1}{2}$ in. dia. $4\frac{1}{8}$ in. long   2 in. of $\frac{3}{16}$ in. CR rod

Procedure for Making the Screw

1. Take the $\frac{1}{2}$ in. CR round; lay out the hole location and punch. Place it in one or two vee-blocks, depending on their size. Lightly clamp with about 1 in. extended from one end.
2. Set up the punch mark so it is centered and on top by using the combination square and a rule. Refer to Section H, Unit 4, Figure 29, page 376 in the text.
3. With a wiggler, locate the punch mark directly under the spindle.
4. Center drill; change to a $\frac{3}{16}$ in. drill and drill through. Chamfer both sides lightly. This part is now ready for the lathe work.
5. Chuck the material in a three-jaw chuck and center drill the end (the end that is not cross drilled) with a centerdrill. (Since this small drill is very delicate, use a high speed and a very light feed.) A number 3 centerdrill will work if you do not go too deeply.
6. Reverse the work in the chuck and machine the radius on the cross drilled end. Check with a radius gage; file to finish the shape of the radius.
7. Again, reverse the work and grip the radius end with the three-jaw chuck for about $\frac{1}{4}$ inch of the stock. Place the dead center in the other end. Use center lube.
8. Machine to size and make the shoulders to length. Turn an undercut for terminating the thread.
9. Cut threads $\frac{3}{8}$-24 to proper depth with a single point tool, not with a die; check with the tapped $\frac{3}{8}$-24 thread in the C-clamp body.

Procedure for Making the Pin

1. Saw off 2 in. of $\frac{3}{16}$ in. CR rod. Make a full radius on each end.
2. Knurl or prick punch at the center to make a tight fit in the $\frac{3}{16}$ in. hole. Insert the rod and tap in place.

WORKSHEET 3. Making the Swivel

Materials

Approximately 2 in. long piece of $\frac{3}{4}$ in. diameter CR round

Procedure for Making the Swivel

1. Set up in a three-jaw chuck.
2. Machine the swivel end as shown on Figure 1 and part it off. Deburr.

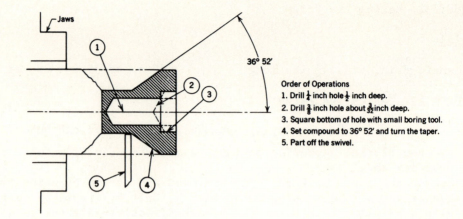

Figure 1

Procedure for Assembling the C-clamp

1. Assemble the clamp screw in the clamp body.
2. Place swivel end on the $\frac{1}{4}$ in. diameter of the screw.
3. Place a $\frac{5}{16}$ in. bearing ball in the hole formed by the centerdrill. (This center-drilled hole must be large enough to leave a very thin edge so it will roll over from the pressure of the bearing ball.) Place a piece of scrap metal on the C-clamp anvil to protect it, and apply pressure by turning the screw. This procedure enlarges the end of the screw, thus retaining the swivel, yet allowing it to turn.
4. Turn in the grading sheet and finished project for your instructor's evaluation.

C-CLAMP: DRAWING I

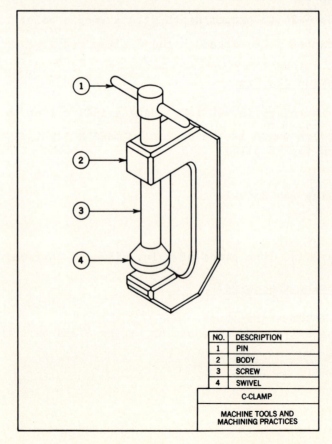

NO.	DESCRIPTION
1	PIN
2	BODY
3	SCREW
4	SWIVEL
	C-CLAMP
	MACHINE TOOLS AND MACHINING PRACTICES

C-CLAMP: DRAWING II

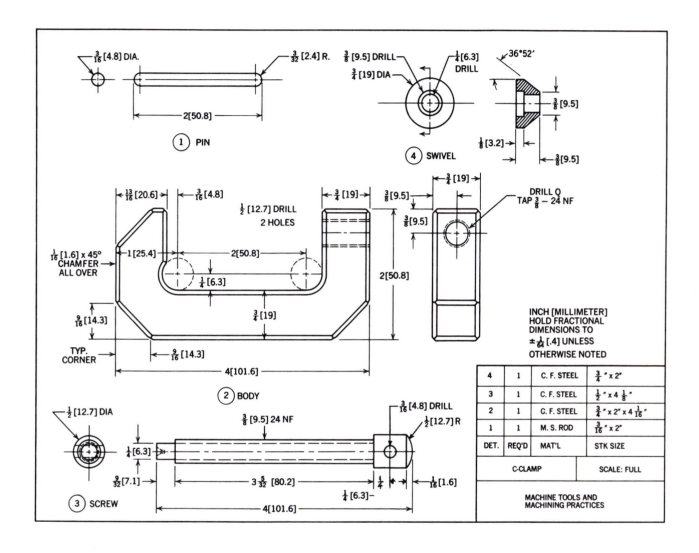

MACHINE TOOL PRACTICES

Name _____ Date _____

PROJECT 5. PAPER PUNCH

Project Evaluation (To be filled out by the instructor):

	Grade	
	Letter	Percent
1. Follows drawing (dimensions and tolerances)	_____	_____
2. Machining finishes	_____	_____
3. Mechanism or tool operates satisfactorily	_____	_____
4. General workmanship	_____	_____
Total grade	_____	_____

Comments:

Signed: _____
(Instructor)

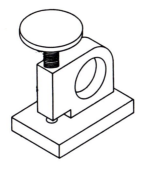

PROJECT 5. PAPER PUNCH

Objectives

    1. Learn to drill and ream holes in alignment.
    2. Be able to tap threads in blind holes.
    3. Learn countersinking and reaming techniques.
    4. Be able to turn a large radius, using a lathe.
    5. Be able to cut a thread with a hand die.
    6. Be able to wind a spring by hand.

Outline for Study

    Prior to starting each procedure for this project, study and complete Post-Tests for:

    1. Making the body: Section H, Units 5 and 6; refer to Unit 4;
                           Section B, Units 5, 7, and 8.
    2. Making the base: Refer to Section H, Units 5 and 6.
    3. Making the punch: Section B, Unit 8.
    4. Hardening the base and punch: Refer to Section D, Unit 3.

Procedures

    Begin the procedures for this project by completing the following worksheets.

Worksheet 1. Making the body and base
Worksheet 2. Countersinking the base
    Exercise 1. Identifying machine reamers
Worksheet 3. Reaming the punch hole and 1 inch hole
    Exercise 1. Identifying hand reamers
Worksheet 4. Making the punch
Worksheet 5. Making the mushroom cap
Worksheet 6. Making the spring
Worksheet 7. Hardening and tempering the punch and base

WORKSHEET 1. Making the Body and Base

Materials

Small countersink $\frac{1}{4}$ in. drill $\frac{3}{16}$ in. drill $\frac{31}{32}$ in. drill
No. 7 drill
Centerdrill $2\frac{1}{16}$ in. of $\frac{1}{2}$ x $1\frac{1}{2}$ in. CR bar

                $2\frac{9}{16}$ in. of $\frac{5}{16}$ x $1\frac{1}{2}$ in. chrome spring steel

Procedure

1. Lay out the body (Figure 1) and the base (Figure 2) and punch mark where holes will be drilled except for the punch guide hole. It will be drilled later when the body and base are bolted together. This procedure will insure perfect alignment of the punch and die hole in the base.

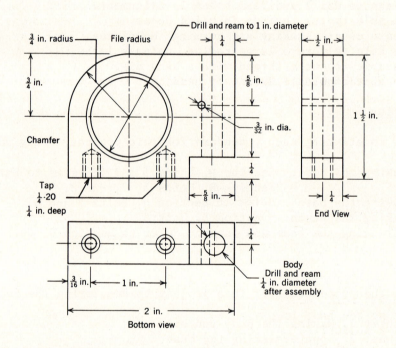

Figure 1.

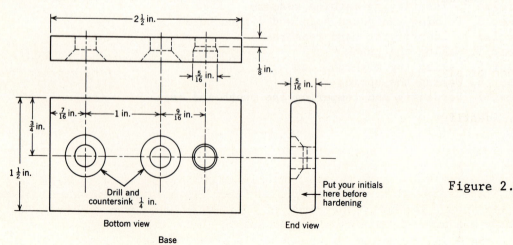

Figure 2.

2. Set up the body (bottom up) in a drill press vise.
3. Spot the center punch mark opposite the punch hole with the centerdrill and set up the No. 7 drill in the chuck. Set the depth stop so the No. 7 drill will not go past the layout line of the one inch hole.
4. Drill to depth. Countersink slightly for tapping. This hole should now be tapped with a $\frac{1}{4}$-20 tap. (Refer to the units on tapping in Section B.) Do not drill and tap the other hole until the base is clamped on.
5. Take the body out of the drill press vise and place the base material in the vise.
6. Drill both bolt holes in the base (not the punch hole) with a No. 7 drill.
7. Now drill out the one that corresponds to the hole that you have just tapped with a $\frac{1}{4}$ in. drill size.
8. Using a $\frac{1}{4}$-20 x $\frac{1}{2}$ in. capscrew for a temporary fastener, bolt the base to the body and align them.
9. Place the assembly in a drill press vise, bottom up.
10. Using the No. 7 drill, set the depth stop and then drill the other hole in the body using the No. 7 hole in the base as a guide. The procedure you have just used to ensure alignment of two holes in an assembly is commonly used by machinists because it is more accurate than separately laying out and drilling two pieces that must fit together.
11. Remove the bolt holding the base and tap the second hole in the body.
12. Drill the second hole in the base to $\frac{1}{4}$ in. dia.

WORKSHEET 2. Countersinking the Base

Material

82 degree countersink at least $\frac{5}{8}$ in. diameter
Cutting oil   Drill press and tooling

Procedure

1. Set the drill press for the lowest speed, as the base is a tool steel that will rapidly work harden at higher speeds.
2. Set up the base in a vise and chuck the countersink.
3. Apply sulfurized cutting oil.
4. Turn on the machine and begin cutting the countersink. Always produce a chip. If the tool rubs, the result will be workhardening of the surface and starting a new cut will be difficult.
5. When you are approaching the correct depth, stop cutting and check the fit with a $\frac{1}{4}$ in. countersink screw by dropping it into the hole. It should be flush with the surface or slightly below (about 0.010 to 0.015 in.). The size may also be checked by using tables of screw head sizes found in handbooks.

EXERCISE 1. Identifying Machine Reamers

Procedure

1. Obtain a prepared kit of reamers or ask your instructor to direct you to the place where reamers are stored, if they are not readily available to you.
2. Practice naming the different kinds of machine reamers until you can easily identify and name any reamer when you see one.
3. Select the $\frac{1}{4}$ in. chucking reamer for the next worksheet.

WORKSHEET 3. Reaming the Punch Hole and the One Inch Hole

Material

One $\frac{1}{4}$ in. diameter fluted chucking reamer   Appropriate drills
One 1 in. diameter shell reamer   Sulfurized cutting oil

Procedure

1. At this point the holes should all have been drilled with the exception of the punch hole.
2. Bolt the body and base together with flat head screws and set up in the drill press (Figure 3).
3. Spot the layout punch mark on the top of the body and drill through the body and base together.
4. Drill through both body and base with a C drill.
5. Maintaining the setup, ream to $\frac{1}{4}$ in. diameter.

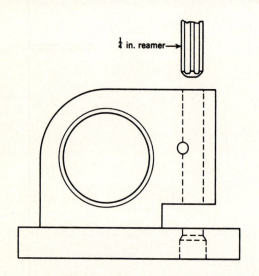

Figure 3.

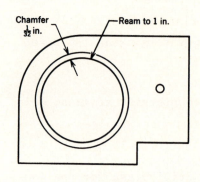

Figure 4.

6. Now drill a $\frac{5}{16}$ in. diameter hole part way into the base from the bottom to allow for paper chaff clearance.
7. Set up and align for cross drilling the $\frac{3}{32}$ in. hole through the $\frac{1}{4}$ in. hole. Be sure to plug the $\frac{1}{4}$ in. hole with a soft steel round. It may be necessary to mark it a few times with a punch to make it tight in the $\frac{1}{4}$ in. hole. Align the spindle with the layout punch mark and chuck a $\frac{3}{32}$ in. drill. Drill through.
8. Using an upright or radial drill press, set up for drilling a $\frac{31}{32}$ in. hole in the body. Spot drill, then use a large pilot drill followed by the $\frac{31}{32}$ in. drill size.
9. Maintain setup and machine ream to 1 in. (Figure 4).
10. Using a large countersink, chamfer the edges of the 1 in. hole $\frac{1}{32}$ in.
11. The notch on the bottom of the body may now be sawn out and filed to the layout line.

EXERCISE 1. Identifying Hand Reamers

1. Ask your instructor to direct you to the location where a variety of reamers are stored. With the aid of this unit, identify the different kinds of hand reamers. Practice this until you can identify and name a hand reamer when you see one.
2. Hand ream the $\frac{1}{4}$ inch diameter punch hole with an expansion reamer so that the punch has a sliding fit.

WORKSHEET 4. Making the Punch

Material

$2\frac{9}{16}$ in. of $\frac{1}{4}$ O-1 or W-1 drill rod

Procedure

1. Saw the $2\frac{9}{16}$ in. length of $\frac{1}{4}$ in. drill rod, using a hand hacksaw and vise jaw caps.
2. Chamfer one end and thread with a hand die, $\frac{1}{4}$-20, $\frac{1}{4}$ in. long. Use sulfurized cutting fluid and back up the die frequently to break up the chips.
3. File the radius on the other end as shown in Figure 5, using a $\frac{5}{16}$ in. diameter round file. Deburr.
4. File a flat, as shown on the drawing, approximately $\frac{3}{64}$ in. deep.
5. Insert the punch into the guide hole and align the flat with the $\frac{3}{32}$ in. guide hole.

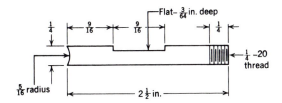

Figure 5.

6. Cut off a $\frac{9}{16}$ in. length of $\frac{3}{32}$ in. rod and insert it in the body, locking the punch in. (Do not peen over the $\frac{3}{16}$ rod yet as the punch has to be removed to harden it.)
7. If the punch does not move freely along the flat, then remove it and file it to a greater depth. Also make sure the cutting end of the punch slides completely back into the body. If it does not, file the flat longer toward the cutting end.
8. The punch should also have a sliding fit (free but not loose) in the hole. If the machine reamer did not provide a sliding fit, then a hand reamer or an expansion and reamer should be used to provide the correct fit.

WORKSHEET 5. Making the Mushroom Cap

Material

$\frac{9}{16}$ in. of $1\frac{1}{2}$ in. dia. HRMS round bar

Procedure

1. In a small lathe, using a three-jaw chuck, set up the $1\frac{1}{2}$ in. diameter piece to turn relatively true (no more than $\frac{1}{32}$ in. axial runout).
2. Face one end and centerdrill. Drill $\frac{3}{8}$ in. deep and tap $\frac{1}{4}$-20, first with a plug tap followed by a bottoming tap (Figure 6).

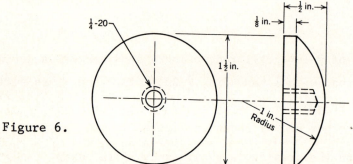

Figure 6.

3. Reverse the cap in the chuck, gripping less than $\frac{1}{8}$ inch.
4. A 1 inch radius may be turned with a radius attachment or it may be done freehand. If it is done freehand, a radius gage should be frequently used to check your progress. Tool marks can be removed by filing, first with a rough or double cut file and then a single cut file. The cap may then be finished with abrasive cloth backing the cloth up with a file so your fingers will be away from the chuck jaws.

WORKSHEET 6. Making the Spring

Materials

$\frac{1}{4}$ in. spring winding mandrel   Two hardwood blocks   No. 17 (.039 in.) music wire

Procedure

1. Use 17 gage spring wire to wind a spring about 1 in. long. Place two blocks of hardwood in the vise. Place a $\frac{1}{4}$ in. mandrel (slotted on the end) between the blocks (Figure 7).
2. Insert the end of the spring wire in the slot. Hold the wire at approximately the lead angle of the spring and turn the mandrel with the vise tight enough to drag on the mandrel. Make a spring about one inch long.
3. When the spring is wound, pull it from the mandrel and cut or break it to length by nicking it on the edge of a grinding wheel. (Spring wire should <u>not</u> be cut with wire cutters.) If the coils are too close together, grip the ends of the spring with pliers and pull it apart until the coil spacing is correct (Figure 8). Flatten the ends of the spring by grinding on the flat side of the wheel.

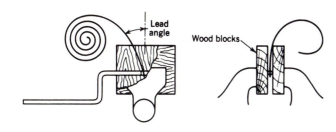
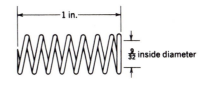

Figure 7.   Figure 8.

WORKSHEET 7. Hardening and Tempering the Punch and Base

Procedure

1. See Section D, Unit 3, before doing any of the heat treating.
2. Heat about $\frac{3}{8}$ in. of the cutting end of the punch with a torch to about 1500°F (a cherry red color) and quench in oil or water, depending on the composition of the steel, whether it is O1 or W1. The reason the punch is not hardened for its full length is the possibility of warping, especially with a flat on one side.
3. Temper the punch on a hot plate.
4. The base may be hardened all over as in a furnace and quenched in the proper medium. Be sure it is chamfered all over (has no sharp edges) except for the punch hole.
5. Polish the base by laying a strip of abrasive cloth on a hard, flat surface and sliding the base back and forth over the abrasive.
6. Temper the base on a hot plate.

Assembly of the Paper Punch (Drawing I)

1. Make sure all parts are chamfered before assembly.
2. After you have assembled the parts of your paper punch, check it for operation Lightly oil the punch. Check to see if it will cut one thickness of paper cleanly.
3. If everything works well, the $\frac{3}{32}$ in. pin may be slightly riveted to retain it in the hole.
4. The pin may now be filed flush with the body.
5. Turn in the grading sheet and the finished paper punch for your instructor's evaluation.

PAPER PUNCH: DRAWING I

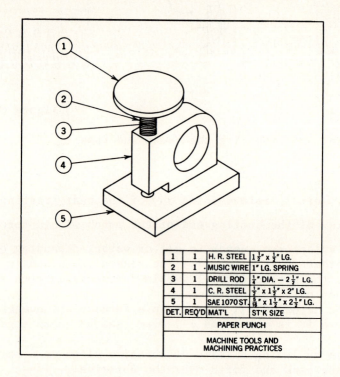

MACHINE TOOL PRACTICE

Name _____   Date _____

PROJECT 6. PARALLEL CLAMP

Project Evaluation (To be filled out by the instructor):

	Grade	
	Letter	Percent
1. Follows drawing (dimensions and tolerances)	_____	_____
2. Machining finishes	_____	_____
3. Mechanism or tool operates satisfactorily	_____	_____
4. General workmanship	_____	_____
Total grade	_____	_____

Comments:

Signed: _____
(Instructor)

MACHINE TOOL PRACTICES   45

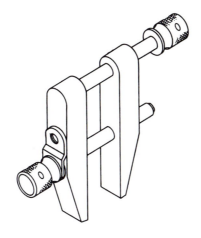

PROJECT 6.   PARALLEL CLAMP

Objectives

   1. Learn to drill two or more holes through two parts and maintain alignment for
      true position.
   2. Learn how to make long threads on slender parts with a button die holder.

Outline for Study

   Prior to starting each procedure for this project, refer to the following:

   1. Making the legs and clip: Section G, Unit 4.
   2. Making the screws: Section I, Unit 9, Figure 37.

Procedures

   Begin the procedures for this project by completing the following worksheets.

   Worksheet 1. Making the Legs and Clip
   Worksheet 2. Making the Screws

WORKSHEET 1. Making the Legs and Clip

Material

   21 tap drill $\frac{1}{4}$ in. dia. drill   Two pieces of $\frac{5}{8}$ x $\frac{5}{8}$ in. keystock 4 in. long
   I tap drill
   P drill $\frac{1}{2}$ in. drill   One piece of $\frac{1}{16}$ in. sheet steel 1 x 2 in.

                      $\frac{3}{16}$ in. drill

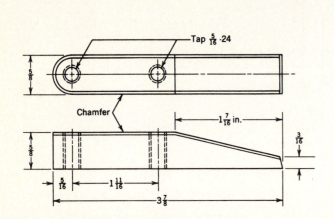

Figure 1.

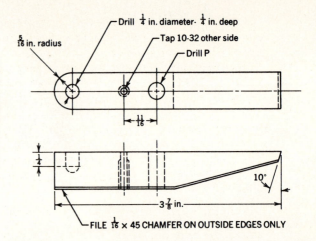

Figure 2.

Procedure for Making the Legs

1. Clamp both legs together with the long flat sides together as they will be when assembled. Set up in the drill press vise.
2. Place the leg up that will not have the 10-32 tapped hole (Figure 1).
3. Spot both layout punch marks. Chuck a $\frac{1}{4}$ in. drill.
4. Set the depth stop so the $\frac{1}{4}$ in. drill will go through one leg and into the other only $\frac{1}{4}$ in. Drill the end hole.
5. Drill the other hole through both legs.
6. Remove the setup and unclamp the legs.
7. Set up the leg with the $\frac{1}{4}$ in. hole that is $\frac{1}{4}$ in. deep (Figure 2). Drill the hole for the 10-32 tap with a No. 21 drill. Drill the center $\frac{1}{4}$ in. hole out to a P drill size.
8. Set up the other leg (Figure 1) and drill it through both holes with an I drill size.
9. Chamfer all holes. Tapping may be done now.
10. Bolt the legs together with the $\frac{1}{4}$ in. hole inside and set up in a bench vise for filing the radius.
11. Rough file both legs together almost to the radius line that you have previously scribed.
12. Finish file, checking often with a $\frac{5}{16}$ in. radius gage.
13. Remove your filing setup when finished and carefully lay the two legs together as they will be when the clamp is assembled. (They should already be bolted together in this position.)
14. Apply layout dye to the sides where you will scribe the angular lines.
15. Scribe the lines. Make sure they are on the correct side of the legs.
16. Saw about $\frac{1}{32}$ in. away from the lines with a hacksaw or on the vertical band saw.

    If you use the band saw, use a push stick and have the correct speed setting for the material.

17. With the two legs clamped together in a bench vise, rough file the angles and then finish file. Your instructor may want you to mill these surfaces.
18. File a 45 degree by $\frac{1}{16}$ in. chamfer on the angles and outer edges only and around the radius. Do not chamfer the clamping side except for lightly deburring it to remove the sharp edge.
19. Case hardening is not recommended as the legs tend to warp when quenched.
20. If gun blueing or machinery black compounds are available in your shop, you may wish to put a finish on the legs.

Procedure for Making the Clip

1. The material used should be larger than the finished size because of the difficulty of holding and drilling the part. The $\frac{1}{2}$ in. hole should be drilled first for this reason.
2. The offset should be made in a bench vise by placing two $\frac{1}{16}$ in. shims on opposite sides of the clip in an offset position. When the vise is tightened, the clip will be offset.
3. Layout and drill the $\frac{3}{16}$ in. hole. The clip may then be finished (Figure 3).

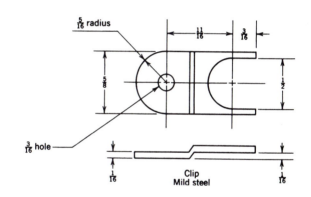

Figure 3.

WORKSHEET 2. Making the Screws

Material

Lathe   Grooving tool   Knurling tool   Parting tool
Hand die
                                 10 in. of CF round stock

Procedure

1. The 10 in. of $\frac{5}{8}$ in. CF round stock will give you sufficient material to make the two screws for the parallel clamp with extra length to hold in the chuck jaws. Use a three-jaw chuck.
2. Centerdrill both ends. When one screw has been finished and parted off, the remaining piece can then be turned end for end. This saves set up time.
3. Extend about 5 in. of stock out from the chuck. Lubricate and adjust the dead center. Begin with the screw with the clip groove (Figure 4).
4. Lay out with a scribe and a rule on the circumference of the work the positions for grooving and knurling.
5. Set up for knurling. Use a medium knurl, starting $3\frac{1}{4}$ in. from the tailstock end. Knurl about $\frac{7}{8}$ in.

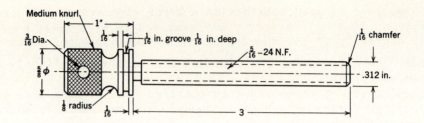

Figure 4.

6. Scribe a mark $3\frac{1}{16}$ in. from the tailstock end and with a $\frac{1}{16}$ in. wide toolbit; make a groove $\frac{1}{16}$ in. deep next to the scribe mark.

7. Make the $\frac{1}{8}$ in. radius groove with a round nose tool. Check often with a radius gage while it is being cut. Use cutting oil.

8. Turn the $\frac{5}{16}$ in. diameter for the length of 3 in. Chamfer.

9. Set up a $\frac{5}{16}$-24 die and die stock, or button die holder if one is available. Use the tailstock spindle as a guide to start the die if a die stock is used. See Section H, Unit 9, page 430 in your textbook; note Figures 36 and 37.

10. Using a low speed, bring the die and holder to the work. Use cutting oil and start the die. When you have made about $\frac{1}{2}$ in. of threads, stop the lathe and back up the holder to break the chips. Repeat this operation until the threads are completed.

11. Set up a parting tool and begin cutting off the screw 4 in. from the end. Part off only near the chuck. When the tool is about $\frac{1}{8}$ in. deep, change to a left-hand roughing tool and make the chamfer $\frac{1}{16}$ in. wide. Replace the parting tool and continue to cut off the screw.

12. Turn the material end for end and chuck about $\frac{3}{4}$ in. of the end. Measure the total depth of the center hole and record it (so you can remove that amount later). Lubricate the center hole and adjust the dead center. You are now ready to proceed with the screw that has the swivel end (Figure 5).

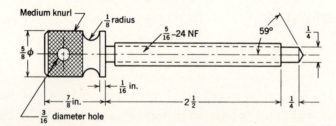

Figure 5.

13. Since the center hole will be removed later on this screw, extra length must be allowed on the threaded portion for this operation. Knurling should begin as near as possible to the chuck and to within 3 in. of the end.

14. The $\frac{1}{8}$ in. radius groove is made as before, after you have done the layout.

15. Turn the $\frac{5}{16}$ in. diameter and the $\frac{1}{4}$ in. diameter on the end.

16. Cut the $\frac{5}{16}$-24 thread with the die.

17. Set up a parting tool and begin to cut off $\frac{7}{8}$ in. from the shoulder of the threaded section. Chamfer as before and complete the cutting off.
18. Screw two $\frac{5}{16}$-24 nuts about an inch onto the thread and jam them tightly together.
19. Mount the screw in a three-jaw chuck on the double nuts with the center hole extending out.
20. Set the compound to 59 degrees with the lathe centerline and turn the swivel end to $\frac{1}{4}$ in. in length. Remove the work from the chuck and turn off the nuts.
21. The $\frac{3}{16}$ in. hole in the knurled knob should be drilled. The knurl must be protected from damage with aluminum or soft copper when it is clamped in a V-block or vise.
22. The parallel clamp may now be assembled.

Assembly of the Parallel Clamp (Drawing I)

1. The parallel clamp is assembled by turning the correct screws in their legs and by fastening the clip with a 10-32 filister head screw $\frac{3}{8}$ in. long.
2. Run the legs of your parallel clamp together. They should come together easily and squarely.
3. Turn in the grading sheet and completed parallel clamp for your instructor's evaluation.

PARALLEL CLAMP: DRAWING I

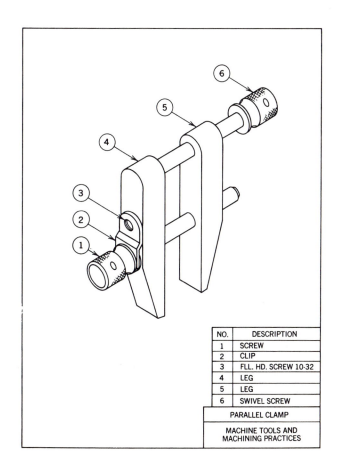

NO.	DESCRIPTION
1	SCREW
2	CLIP
3	FLL. HD. SCREW 10-32
4	LEG
5	LEG
6	SWIVEL SCREW

PARALLEL CLAMP

MACHINE TOOLS AND MACHINING PRACTICES

MACHINE TOOL PRACTICES

Name _____   Date _____

PROJECT 7. TAP HANDLE

Project Evaluation (To be filled out by the instructor):

	Grade	
	Letter	Percent
1. Follows drawing (dimensions and tolerances)	_____	_____
2. Machining finishes	_____	_____
3. Mechanism or tool operates satisfactorily	_____	_____
4. General workmanship	_____	_____
Total grade	_____	_____

Comments:

Signed: _____
(Instructor)

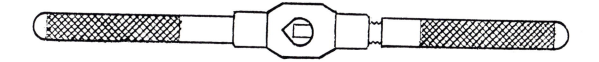

PROJECT 7. TAP HANDLE

Objectives

1. Learn to set up and turn rectangular stock.
2. Learn to tap threads on the lathe.
3. Learn to knurl slender stock.

Outline for Study

Prior to starting each procedure for this project, study and complete Post-Tests for:

1. Making part Number 1: Section C, Units 5 and 7.
2. Making part Number 2: Refer to Section I, Units 6, 11, and 12.

Procedures

Begin the procedures for this project by completing the following worksheets.

Worksheet 1. Making Part Number 1
Worksheet 2. Making Part Number 2

Refer to the drawing at the end of this worksheet, Tap Handle: Drawing I.

WORKSHEET 1. Making Part Number 1

Material

$5\frac{1}{2}$ in. of HRMS flat bar $\frac{1}{2}$ x 1 in.

Procedure

1. Cut off a piece of hot rolled mild steel (HRMS) flat bar $\frac{1}{2}$ x 1 x $5\frac{1}{2}$ in. long.
2. Clean the lathe spindle nose and mount a four-jaw chuck on it.
3. Fasten your workpiece in the chuck and let approximately 1 in. extend out from it.
4. Center your work to within .003 in., using methods shown in Section I, Unit 6, Figures 11 and 12.
5. Use a right-hand turning tool and face the end with a good finish. Do not remove more material than necessary and do not leave a stub in the center.
6. Mount a No. 3 centerdrill in the tailstock drill chuck.
7. Select and set the speed for centerdrilling. If the speed is too slow, the centerdrill will break.
8. Centerdrill the end. Use cutting oil.

9. Take the workpiece out of the chuck, reverse it, and repeat steps 4 through 8.
10. Remove the workpiece from the chuck.
11. Remove the chuck and substitute a drive plate and center in the headstock.
12. Fasten the dog on your workpiece and set it up for turning between centers.
13. Select the speed for roughing.
14. Select the feed for roughing.
15. Mark your workpiece 2 in. from the tailstock end and take roughing cuts to this mark until the diameter is approximately .800 in.
16. Now mark your workpiece $\frac{7}{16}$ in. from the tailstock end and turn that portion to approximately .560 in.
17. Reverse your workpiece end for end, clamping the dog on the end you have just turned.
18. Mark your workpiece $1\frac{9}{16}$ in. from the headstock end and rough turn to that mark until the diameter is approximately .560 in.
19. Now make a mark $2\frac{1}{6}$ in. from the headstock end and turn your piece to approximately .430 in dia. to that mark. Part No. 1 is now all machined to oversize dimensions and is ready for the finishing operation.
20. Resharpen your tool so it can be used for finishing, or make a finishing tool.
21. Select a speed for finishing. This will be slightly higher than for roughing.
22. Select a feed for finishing. This will be much lower than that used for roughing.
23. Mark the work at 2 in. from the headstock end.
24. Finish turn the $\frac{3}{8}$ in. diameter with a tolerance of .377 to .373 in. Machine to the mark.
25. Finish turn the $\frac{1}{2}$ in. diameter with a tolerance of .500 to .494 in. It should end exactly $1\frac{1}{2}$ in. from the headstock end.
26. Set up a medium diamond knurling tool in the tool post.
27. Use a relatively low speed and a feed of about .010 to .020 in.
28. Make a mark $2\frac{13}{16}$ in. from the headstock end; this is how far your knurl will go.
29. Bring the knurls to the work near the dead center and feed in until you can see a definite diamond pattern develop. If there is one solid line one direction and short broken lines in the other, either raise the tool up, lower it, or turn it slightly sideways so it will "bite" in better. If that does not work, find a different knurling tool and try that. Use lube or cutting oil on the knurls and on the work. When a full diamond pattern develops, stop knurling, since further working will ruin the job. Two passes should be sufficient; however, this workpiece will spring in the middle so more passes may be necessary at the center of the workpiece. Remove the knurls while the work rotates to avoid bending the work.
30. Reverse your workpiece; protect the knurled section with some soft material such as aluminum when you put the dog on.
31. Change speed and feed for a finishing cut and set up your right-hand turning tool again.
32. Turn the large diameter with a tolerance of .750 to .748 in.
33. Turn the $\frac{1}{2}$ in. diameter nearest the tailstock end with a tolerance of .500 to .494 in. and $\frac{1}{2}$ in. long.
34. Set the compound rest so it will cut the 30 degree angle; also set up the tool for the cut (Figure 1).
35. Lock the carriage in place and cut the taper. Make sure you get a smooth transition between the $\frac{1}{2}$ in. diameter and the taper.

Figure 1.

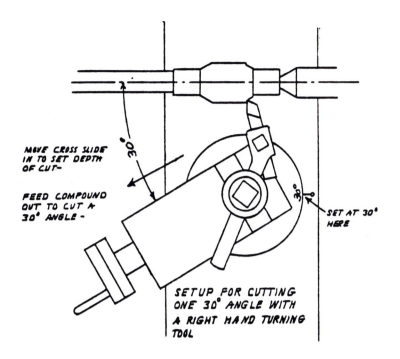

Figure 2.

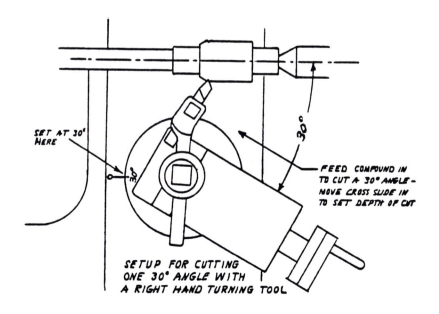

36. Swivel the compound rest to cut the other 30 degree bevel (Figure 2).
37. Turn this taper, being careful also to make a smooth transition between the taper and $\frac{1}{2}$ in. diameter.
38. Set the cutting edge of the tool at 45 degrees to the work and make the two required chamfers.
39. Remove the dog from the workpiece and change to a four-jaw chuck on the spindle.
40. Mark your workpiece 5 in. long. Use a protective aluminum sleeve and chuck the workpiece with the small end out. Allow approximately 1 in. to extend from the chuck.
41. Adjust the chuck jaws until the workpiece runs true.
42. Face off the excess material (to your mark).
43. Rough turn the $\frac{3}{16}$ in. radius and use a file to complete the shape and to finish it. Check the radius with a gage.
44. Remove your workpiece from the chuck.
45. Lay out for the $\frac{3}{8}$ drilled hole on the flat side of the large diameter. Center punch.
46. Clamp the large diameter in a drill press vise, using soft protective material on the jaws. Use a level on the part to make sure it is set up square to the spindle <u>after the vise is tightened</u>. Check to see if the drill press table is level; if not, use the same bubble position for setting up the part.
47. Spot drill with a small centerdrill.
48. Drill the $\frac{3}{8}$ in. hole and chamfer lightly on both sides with a countersink.
49. Chuck the workpiece on the $\frac{3}{8}$ in. diameter using the protective sleeve; this time let the large part extend out from the chuck.
50. Adjust the jaws until the work runs concentric.
51. Find the size of the tap drill for $\frac{5}{16}$-24 NF threads and secure a drill.
52. Hold this drill in the tailstock drill chuck and drill the part to the depth of one inch. The centerdrilled hole will act as a pilot to start the tap drill.
53. Remove the drill and fasten a $\frac{5}{16}$-24 NF tap in the drill stock.
54. Loosen the tailstock clamp so it can slide freely on the ways.
55. Put cutting oil on the tap and turn the chuck with the workpiece in it by hand. After you bring the tap in contact with the work, it will pull itself into the work. Back up the tap frequently to break up the chips.
56. When the tap has reached full depth, turn the chuck in reverse by hand. This will push the tap out.

WORKSHEET 2. Making Part Number 2

<u>Material</u>

$4\frac{3}{4}$ in. of $\frac{3}{8}$ in. dia. CR round stock

<u>Procedure</u>

1. Cut off a piece of $\frac{3}{8}$ in. diameter cold rolled round stock $4\frac{3}{4}$ in. long.
2. Set up your workpiece in a three-jaw chuck leaving 1 in. extending from the chuck. Check it for runout. If it is excessive, use a four-jaw chuck.
3. Center drill the workpiece. Then extend it 3 in. from the chuck.

4. Support this end with a center in the tailstock. Use center lube.
5. Put a mark on your workpiece $2\frac{9}{16}$ in. from the tailstock end. This is how far the knurl will extend.
6. Repeat steps 26 to 29 in the procedure for making part No. 1.
7. Use a protective sleeve to hold the knurled piece so that the end with the center hole extends 1 in. from the chuck.
8. Use your right-hand tool and cut $\frac{3}{8}$ in. off the end of the workpiece. This should remove the center hole.
9. Rough turn and use a file to make the $\frac{3}{16}$ in. radius.
10. Remove your workpiece from the chuck and mark it $2\frac{11}{16}$ in. from the rounded end.
11. Chuck the workpiece with protective material over the knurl so your mark is $\frac{1}{2}$ in. from the chuck.
12. Center drill the end and support it with a center.
13. Turn the $\frac{5}{16}$ in. diameter with a tolerance of .312 to .310 in. diameter to the mark.
14. Turn the $\frac{1}{4}$ in. diameter at the end with a tolerance of .250 to .248 in. diameter so that 1 in. remains of the $\frac{5}{16}$ in. diameter.
15. Make a small undercut $\frac{3}{32}$ in. wide to the minor diameter of your thread.
16. Set up your threading tool and thread the workpiece. Make sure your tool is sharp. (An alternate method is to use a button die holder in the tailstock; no undercut is needed in that case.)
17. Use your right-hand tool to cut the $\frac{1}{4}$ in. diameter back so it will be $\frac{5}{16}$ in. long.

    If an aluminum protective sleeve is used on the knurl, the end may be turned off while set up as a chucking operation (without the dead center).
18. Chamfer the $\frac{1}{4}$ in. diameter to the specifications on Drawing I. This completes the turning of Part 2.

<u>Assembling the Tap Handle</u>

1. Use a small square file and make the vee in Part 1. It should be symetrical to the center line.
2. The short $\frac{1}{4}$ in. end may be case hardened if desired in order to prevent it from splaying out with repeated use. This can be done with an acetylene torch. Only the small end is heated to a bright cherry red and then placed in Kasenite. It is reheated and quenched in water.
3. Assemble both pieces. Use oil as a rust preventative.
4. Check to see if it will hold a tap square to the handle.
5. Turn in the grading sheet and the finished tap handle for your Instructor's evaluation.

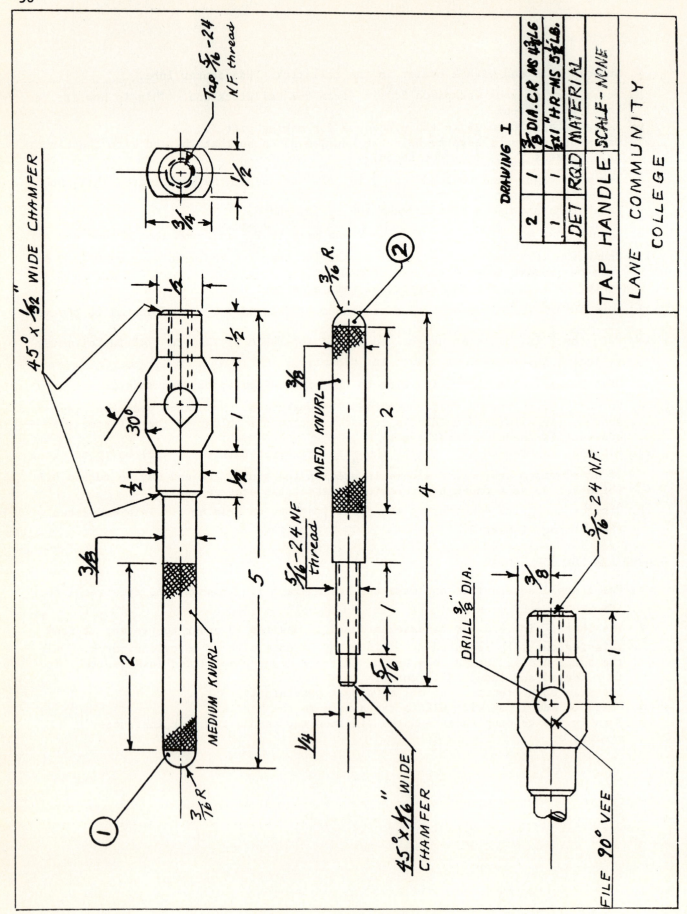

MACHINE TOOL PRACTICES   59

Name _____ Date _____

PROJECT 8. DIE STOCK

Project Evaluation (To be filled out by the Instructor):

                                                                                            Grade

	Letter	Percent
1. Follows drawing (dimensions and tolerances)	_____	_____
2. Machining finishes	_____	_____
3. Mechanism or tool operates satisfactorily	_____	_____
4. General workmanship	_____	_____
Total grade	_____	_____

Comments:

                                                  Signed: _____
                                                                    (Instructor)

MACHINE TOOL PRACTICES   61

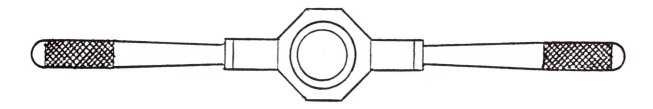

PROJECT 8. DIE STOCK

Objectives

1. Learn how to set up rectangular stock in a four-jaw chuck for boring operations.
2. Bore and counterbore precision internal diameters.
3. Learn to use the taper attachment and perform taper calculations.

Outline for Study

Prior to starting each procedure for this project, study and complete Post-Tests for:

1. Making the body: Section I, Unit 8, and refer to Section I, Unit 9.
2. Making the handles: Refer to Section I, Units 10, 11, and 13.

Procedures

Begin the procedures for this project by completing the following worksheets.

Worksheet 1. Testing Alignment of Centers with a Test Bar
Worksheet 2. Testing for Taper by Turning
Worksheet 3. Making the Body
   Exercise 1. Using Measuring Tools
Worksheet 4. Making the Handles
   Exercise 1. Making a Taper

WORKSHEET 1. Testing Alignment of Centers with a Test Bar

Materials

   Lathe   Test bar   Dial indicator

Procedure

Set up and test for taper between centers in a lathe and adjust the tailstock.

1. Clean the headstock and tailstock spindle tapers and insert the centers.
2. Place the test bar between centers and adjust the lathe centers so there is no end play.
3. Set up a dial indicator in the tool post and adjust the contact point to the height of the centerline of the lathe.
4. Move the dial indicator to the test diameter near the headstock. Zero the bezel.
5. Now move the indicator to the test diameter near the tailstock and observe the reading on the dial. If it is not off more than 0.0005 to 0.001 in., leave it as it is.

6. If the reading is positive (clockwise), move the tailstock away from the operator side until the needle points to zero index.
7. If the reading is negative (counterclockwise), move the tailstock toward the operator until the needle points to zero index.
8. Repeat the test for a double check of your results.

WORKSHEET 2. Testing for Taper by Turning

Material

    Lathe   Micrometer   Test piece
    Dial indicator   Right-hand turning tool

Procedure

Set up and test for taper between centers in a lathe and adjust the tailstock if needed.

1. Clean the headstock and tailstock spindle tapers and insert the centers.
2. Place the test piece between centers with a drive dog. Use high pressure lubricant on the center hole toward the tailstock. Adjust for turning.
3. Set up your right-hand turning tool First, take a roughing cut if the test piece is irregular. Then take a light cut (about 0.015 in. dia.) with a finishing feed as far as you can safely go across the test piece.
4. Measure both ends of the piece and subtract the smaller diameter from the larger. If the difference is not more than 0.0005 to 0.001 inch per foot length, the tailstock should not be changed.
5. Set up the dial indicator to contact the centerline of the end of the test piece at the tailstock end. Zero the bezel.
6. If the tailstock end is smaller in diameter, then move the tailstock away from the operator's side by half the amount of difference in the two diameters.
7. If the tailstock end is larger in diameter, then move the tailstock towards the operator's side by half the difference in diameters.
8. Take another light cut across the test piece. Measure both ends to double check for alignment of the centers.

WORKSHEET 3. Making the Body

Materials

$4\frac{3}{16}$ inches of $\frac{3}{4}$ x 2 in. CR flat bar   Two $\frac{1}{4}$-20 socket head set screws $\frac{1}{4}$ in. long

EXERCISE 1. Using Measuring Tools

1. Inspect your shop and tool crib and identify as many comparison measuring tools as possible.
2. Obtain the small hole and telescoping gage measuring kit from your instructor and note how they work for transferring measurements.
3. If you have a cylindrical or micrometer square available, use it to check some of the combination squares in your shop. The results may surprise you.

Procedure for Making the Body (DRAWING I)

1. Place a bar of $\frac{3}{4}$ x 2 in. cold rolled mild steel in the power saw and cut off $4\frac{1}{8}$ in. plus $\frac{1}{16}$ in. for finishing, or $4\frac{3}{16}$ in. total.
2. Deburr the sharp edges and clean the piece in solvent.
3. Find the center of the large flat side and center punch.
4. Set up in a four-jaw chuck with the center punched side out. Two of the jaws may have to be reversed. The punch mark should be adjusted to run true with the machine axis. A wiggler and a dial indicator may be used for this or a sharp dead center in the tailstock could be used to locate the punch mark. True up the face using a dial indicator.
5. Drill through the material, using first a center drill followed by a pilot drill and a $\frac{3}{4}$ in. diameter drill.
6. Set up a boring toolholder with a $\frac{1}{2}$ in. bar. Set the bar to hold the toolbit in an angular position. Grind the tool to cut properly at that angle so that an internal square shoulder can be made. The bar should extend no further from its holder than necessary.
7. Using the correct RPM and feed, bore to $1\frac{3}{16}$ in. dia. with a tolerance of plus or minus .005 in.
8. Counterbore to a rough size of about .025 in. under finish diameter. Leave $\frac{1}{32}$ in. approximately in the bottom of the counterbore for finish; that is, make the length of the counterbore $\frac{15}{32}$ in.
9. Check your toolbit for builtup edge or dullness and hone if necessary.
10. Set the feed and RPM for finishing.
11. One method of turning a counterbore to depth is to touch the toolbit to the outside surface of the workpiece near the bore. Using a steel rule laid flat on the ways and touching the carriage, mark at an inch line or at the end of the rule on the lathe way with a grease or graphite pencil. (Note: These markers will not harm a precision way, but do not use a scriber or any hard or sharp marking device.) This will provide an index position for moving the carriage. Position the boring bar with the cross slide so the bar and tool can enter the bore. Another method is to use the compound micrometer collar to set the depth of the tool with the compound parallel to the center axis of the lathe.
12. Bring the tool to the inner edge of the bottom of the counterbore and set it to $\frac{1}{2}$ in. depth. Lock the carriage.
13. Start the lathe and slowly hand feed the cross slide to finish the face to depth. Watch the tool in the bore, and when the cut is at the diameter of the large bore, stop feeding and turn off the lathe.
14. Unlock the carriage and carefully clear the tool and back it out from the work.
15. Take a light trial cut on the $1\frac{1}{2}$ in. dia. bore (about 0.005 in.) for a distance of $\frac{1}{8}$ in.
16. Measure the trial cut and subtract the measurement from 1.505 in. The difference will be the depth of the finish cut as measured on the diameter.
17. If the lathe you are using is graduated on the micrometer collar to read on the diameter of the work, then simply set the number you have found in step 16. If, on the other hand, the dial is graduated for single depth, then set the depth of cut at half the value found in step 16.
18. Take another trial cut, this time for about $\frac{1}{4}$ in. Measure and correct any error caused by tool spring. The error will generally be undersize.

19. Finish bore to full depth. Stop the power feed before the tool comes to the bottom and hand feed to make a sharp corner.
20. Chamfer all sharp edges with a tool set at 45 degrees. Remove the workpiece from the lathe.
21. Lay out for center on both ends of the material. Center punch.
22. Center drill in a drill press on both ends, or use an alternate method: Center drill in the lathe and face to length.
23. Tap drill both holes for set screws.
24. Set up for turning between centers. Alternately, you could use a four-jaw chuck and dead center.
25. Rough out both ends to $\frac{21}{32}$ in. dia. by $\frac{31}{32}$ in. long
26. Turn the center section to $1\frac{7}{8}$ in. diameter using a finishing tool and feed for the last cut.
27. Set the compound for 45 degrees and rough turn both angular surfaces to within $\frac{1}{32}$ in. of finish size.
28. Face both ends so that the total length is $4\frac{1}{8}$ inches.
29. Finish turn the $\frac{5}{8}$ in. dia. ends to finish diameter and to the 1 inch length.
30. Finish turn the angular surface with the compound set at 45 degrees. Turn the compound hand crank slowly and evenly to provide a good finish.
31. Remove the part from the lathe and install a three-jaw chuck.
32. Place the $\frac{5}{8}$ in. end in the chuck and tighten.
33. Tap drill $\frac{3}{4}$ in. deep for a $\frac{7}{16}$-20 tap. Chamfer the hole.
34. Place tap in the Jacobs chuck and start by turning the three-jaw chuck by hand. Leave the tailstock loose on the ways. Use cutting oil on the tap. Continue to alternately turn the chuck forward and reverse to break the chips until the tap offers more than usual resistance, showing that it has reached the bottom of the hole. Back out the tap and repeat the process on the other end.
35. Remove the part from the lathe and tap the two holes for the set screws. Clean up the part and deburr where necessary. The body is now completed.

WORKSHEET 4. Making the Handles

Material

10 inches of $\frac{5}{8}$ in. dia. CR shaft

Procedure

1. Place some $\frac{5}{8}$ in. CR round stock in the saw. Cut off two pieces $5\frac{1}{2}$ in. long.
2. Set up material in a three-jaw chuck. Face off and center drill both ends.
3. With one end in the chuck and the other supported with a dead center, knurl from the tailstock end to a mark $3\frac{3}{16}$ in. from the headstock end
4. Machine the step and shoulder for the $\frac{7}{16}$-20 thread.
5. Thread the end by single point tool method. Do not use a die. Check with a $\frac{7}{16}$-20 nut or with the body of the die stock.
6. Machine the taper using the taper attachment. You will find the formula for the various methods of producing a taper in your textbook in Section I, Unit 13. Determine the taper per foot and set that amount on the taper attachment.

7. Chuck the knurled end extending the end opposite the thread, using soft aluminum or copper to protect the knurl. Face the part to length.
8. Freehand cut a $\frac{5}{16}$ in. radius. Check frequently with a radius gage. Finish with a single cut file.
9. Repeat steps 1 through 8 to make the other handle.
10. Clean all parts in solvent and wipe dry. The die stock is now ready for assembly.

EXERCISE 1. Making a Taper

Note: This exercise may be used to make the tapers on the handles if a taper attachment is not available and if you have your instructor's permission.

Material

Lathe and tooling $5\frac{1}{8}$ in. of $\frac{3}{4}$ inch CF round stock

Procedure

(a) Make a taper plug by the offset tailstock method (Figure 1).

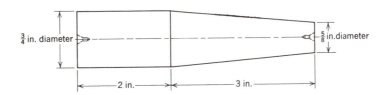

Figure 1.

(b) Measure the taper per inch.

1. Cut off a piece of CF mild steel $\frac{3}{4}$ by $5\frac{1}{8}$ in. long
2. Center drill and face both ends to an overall length of 5 in.
3. Fasten a lathe dog to your workpiece and set up for turning between centers.
4. Calculate the tailstock set over and record here: _____
5. Use a dial indicator or the compound rest slide to offset the tailstock.
6. Take some trial cuts until the tapered portion is approximately 2 in. long.
7. Calculate the taper per inch. Record here: _____
8. Make two marks on the tapered part exactly 1 in. apart with a scriber.
9. Measure the diameters on these marks ans subtract the smaller from the larger. This difference should be the same as the answer in step 7.
10. If it is not the same or within 0.001 in., adjust the tailstock and make trial cuts until the result is correct.
11. Mark the workpiece 3 in. from the tailstock end and make as many passes as necessary to bring the taper to that length.

Assemble the Die Stock

1. Turn in the grading sheet and finished die stock for your Instructor's evaluation.

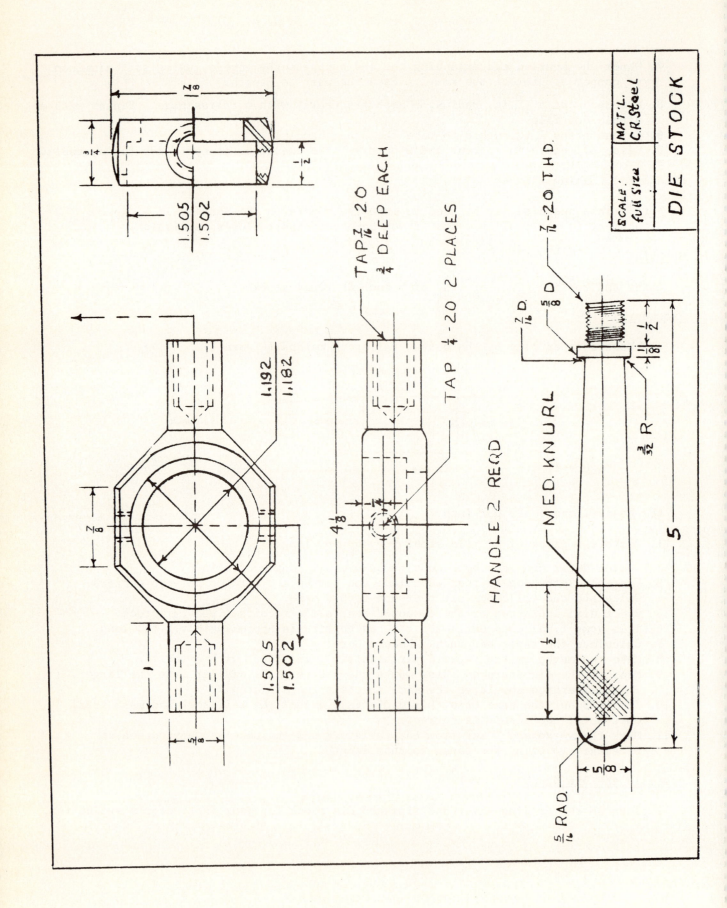

MACHINE TOOL PRACTICES 67

Name _____ Date _____

PROJECT 9. CENTER AND TAPER SLEEVE

<u>Project Evaluation</u> (To be filled out by the instructor):

                                                                                  Grade

	Letter	Percent
1. Follows drawing (dimensions and tolerances)	_____	_____
2. Machining finishes	_____	_____
3. Mechanism or tool operates satisfactorily	_____	_____
4. General workmanship	_____	_____
Total grade	_____	_____

<u>Comments:</u>

Signed: _____

(Instructor)

MACHINE TOOL PRACTICES   69

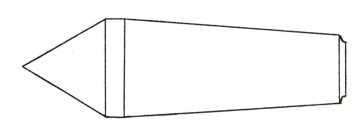 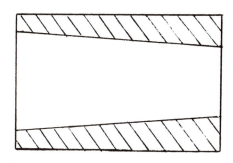

PROJECT 9. CENTER AND TAPER SLEEVE

Objectives

1. Learn how to produce a precision Morse taper.
2. Learn how to make a precision tapered bore.
3. Be able to fit up an external and internal taper to an acceptable degree of accuracy.

Outline for Study

Prior to starting each procedure for this project, refer to the following:

1. Making a Morse taper: Section I, Unit 13.

Procedures

Begin the procedures for this project by completing the following worksheet.

Worksheet 1. Making a Morse Taper
    Exercise 1. Turning a Morse Taper Live Center
    Exercise 2. Turning an Internal Morse Taper Sleeve

WORKSHEET 1. Making a Morse Taper

EXERCISE 1. Turning a Morse Taper Live Center

Make a Morse taper live center (Figure 1). Your instructor may wish to assign you a certain size Morse taper.

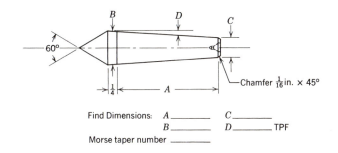

Figure 1.

Find Dimensions: A_____ C_____
                 B_____ D_____ TPF
Morse taper number _____

Materials

One piece of HRMS round stock, 1 in. dia. by $4\frac{1}{2}$ in. long for a No. 3 Morse taper

Other taper sizes will have different dimensions that can be determined by the tables in Section I, Unit 13.

Procedure

1. Cut off mild steel round stock slightly larger in diameter than the B dimension and $\frac{1}{4}$ in. longer than the finished dimension of the center. The length of the 60 degree angle can be determined by multiplying .866 by the large diameter of the center.
2. Set up the stock in a three-jaw chuck and center drill the end.
3. Extend the work outward, holding only $\frac{1}{2}$ in. in the chuck jaws. Support the other end with the tailstock center.
4. Turn the workpiece to diameter B for more than length A plus $\frac{1}{4}$ in.
5. Determine the taper per foot D, and record here: _____
6. Clean and lubricate the taper attachment.
7. Adjust the taper attachment to the correct TPF.
8. Move the carriage so the tool is where the taper is to be cut on the workpiece and lock the anchor bracket to the lathe bed.
9. Disengage the cross feed nut by removing the screw on the cross slide. Cover the hole to keep chips out. This procedure is not necessary if your lathe is equipped with the telescoping type of taper attachment.
10. Tighten the binder handle on the slotted extension of the cross slide.
11. Mark the approximate length of the taper on your workpiece.
12. Move the carriage to make sure that your tool clears the starting end of the taper to be cut by at least $\frac{3}{4}$ in. If you don't have enough travel, adjust the location of the anchor bracket.
13. Select and set your speed and feed.
14. Make a trial cut $\frac{1}{16}$ in. deep and let it cut until it runs out of the work. Start the feed about $\frac{1}{2}$ in. before the beginning of the cut to remove any play in the taper attachment.
15. Take a second, lighter cut 0.020 in. deep so that a measurement can be taken.
16. With a pencil or scribe, make 2 lines exactly 1 in. apart on the tapered part.
17. Determine the correct taper per inch for your center.

    $$TPI = \frac{taper\ per\ foot}{12}$$. Record your answer here: _____

18. Now use a micrometer to measure your workpiece on the two lines; the difference between these diameters should be the same as the taper per inch of your center, as listed in the tables.
19. If the taper is not correct, readjust the taper attachment. Then take trial cuts and measurements until it is correct.
20. Take additional cuts until a finishing cut makes dimension C correct. It should be a very smooth finish.
21. Use a taper ring gage to check for exact fit before getting to final size.
22. Chamfer the end.
23. Remove your workpiece from the chuck and remove the chuck from the lathe.
24. Mount the workpiece in the lathe spindle nose. If necessary, use a sleeve to match the spindle nose taper.

25. Set the compound rest to 30 degrees.
26. Turn the 60 degree taper until the cylindrical part is $\frac{1}{4}$ in. long.
27. This completes the turning of the lathe center. Save it to be used as a test plug for the sleeve and later to be turned in for grading.

EXERCISE 2. Turning an Internal Morse Taper Sleeve

Make a Morse taper sleeve to fit the live center.

Materials

$1\frac{1}{4}$ x $2\frac{5}{8}$ in. HRMS for a No. 3 Morse taper

Procedure

1. Select the diameter of internal taper that corresponds to the taper number of the lathe center you have made.
2. Cut off material $\frac{1}{8}$ in. longer than the workpiece (Figure 2).

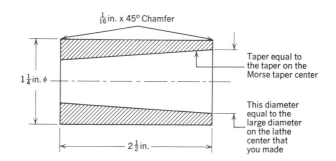

Figure 2.

3. Mount work in a three-jaw chuck. Face and drill the center.
4. Use a drill $\frac{1}{16}$ in. smaller in diameter than the small diameter of the taper sleeve and drill completely through it.
5. Determine the taper per foot for your taper. Record here: _____
6. Wipe the taper attachment clean and lubricate it.
7. Adjust the taper attachment to the required taper.
8. Mount a boring bar in your tool post; the cutting edge must be exactly on center. Check the tool to see that it has enough clearance.
9. Lock the anchor bracket to the lathe bed; move the carriage by hand to make sure you have sufficient travel.
10. Disengage the cross feed nut by removing the screw on the cross slide. Cover the hole to keep chips out. This procedure is not necessary if your lathe is equipped with the telescoping taper attachment.
11. Tighten the binder handle on the slotted extension of the cross slide.
12. Select and set your speed and feed.
13. Make trial cuts until the taper extends part of the way through the bore.
14. Use the lathe center that you made for a plug gage to check the internal taper.
15. On the last few passes use a fine feed of .003 to .004 in. for a smooth finish.
16. When the large diameter of the internal taper is the same diameter as the large diameter on the lathe center, the taper is finished. Chamfer the sharp edges.
17. Turn in the grading sheet and finished project to your instructor for evaluation.

MACHINE TOOL PRACTICES

Name _____   Date _____

PROJECT 10. INTERNAL THREAD

<u>Project Evaluation</u> (To be filled out by the instructor):

	Grade	
	Letter	Percent
1. Follows drawing (dimensions and tolerances)	_____	_____
2. Machining finishes	_____	_____
3. Mechanism or tool operates satisfactorily	_____	_____
4. General workmanship	_____	_____
Total grade	_____	_____

<u>Comments</u>:

Signed: _____
(Instructor)

MACHINE TOOL PRACTICES 75

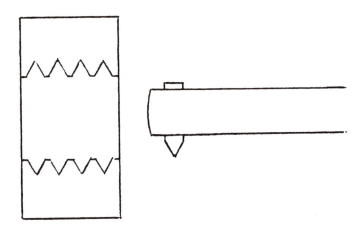

PROJECT 10. INTERNAL THREAD

Objective

Learn how to produce an internal thread by single point lathe threading procedures.

Outline for Study

Prior to starting this project, study and complete the Post-Test for:

1. Cutting an internal thread with a single point tool on the lathe: Section I, Unit 12

Procedures

Begin the procedure for this project by completing the following worksheet:

Worksheet 1. Cutting an Internal Thread on a Lathe

WORKSHEET 1. Cutting an Internal Thread

Material

One inch length of 1½ in. dia. hexagonal or round stock
A precision 1-8 thread plug gage, or a shop-made thread plug gage, or a standard 1-8 UNC capscrew

Procedure

1. Chuck the nut blank and make a facing cut. Chamfer.
2. Calculate the minor diameter for the 1-8 UNC thread. The formula is:

$$d = D - (P \times .541 \times 2)$$

when

P = pitch - $\frac{1}{8}$ in. = .125 in.
d = 1 in. - .125 in. x .541 x 2
d = .86475 in.   or   .865

3. Spot drill the nut blank and drill through with a $\frac{3}{4}$ to $\frac{13}{16}$ in. drill
4. Set up a boring bar and tool; bore the hole to .865 in., the minor diameter of the thread.
5. Set up your lathe for threading 8 TPI and place a threading tool in the boring bar.
6. Swivel the compound rest to 29 degrees toward the chuck.
7. Align the tool with the center gage. Check to see that the chuck jaws will clear the compound when the tool is moved through the bore.
8. Take the slack out of the compound screw by turning it outward and set the tool to the work. Zero both micrometer collars.
9. Take a scratch cut and check the threads with a screw pitch gage.
10. Select a slow speed for threading. Use cutting oil.
11. Determine the infeed on the compound for a 1-8 UNC Unified thread. The formula is:

    Infeed = P x .708

    = .125 in. x .708

    = .088 in.
12. Advance the compound before each cut .005 to .010 in. depending on the rigidity of your setup. Reduce the infeed to .002 in. or less near the last few cuts.
13. When the total infeed depth has been reached, clean out all the chips from the threads and try the thread plug gage for fit. If it will not go in, run the tool through one or two times without advancing the compound slide and try the gage again. If the gage still does not fit, it indicates that the tool form, flat, or alignment is not correct, or that something on your setup has slipped out of place, or that tool spring has not allowed a full depth of cut. Sometimes, several free cuts can be taken to enlarge the diameter. If you are not getting results, continue advancing the tool .002 in. per pass and check with the thread plug gage between each cut.
14. Chamfer the thread and reverse the workpiece in the chuck.
15. Face this side of the nut to length and chamfer both the thread and the outside diameter.
16. Clean up your project in solvent.
17. Turn in the grading sheet and the finished project to your instructor for evaluation.

MACHINE TOOL PRACTICES

Name _____ Date _____

PROJECT 11. MULTIPLE LEAD THREADS

Project Evaluation (To be filled out by the instructor):

	Grade	
	Letter	Percent
1. Follows drawing (dimensions and tolerances)	_____	_____
2. Machining finishes	_____	_____
3. Mechanism or tool operates satisfactorily	_____	_____
4. General workmanship	_____	_____
Total grade	_____	_____

Comments:

Signed: _____

(Instructor)

MACHINE TOOL PRACTICES

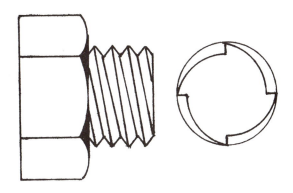

PROJECT 11. MULTIPLE LEAD THREADS

Objectives

1. Be able to produce a two-start external thread on a lathe.
2. Be able to produce a two-start internal thread on a lathe.

Outline for Study

Prior to starting each procedure for this project, study and complete Post-Tests for:

1. Cutting the external two lead thread: Section I, Unit 15
2. Cutting the internal two lead thread: Refer to Section I, Unit 12

Procedures

Begin the procedures for this project by completing the following worksheets.

Worksheet 1. Cutting the External, Two Lead Thread
Worksheet 2. Cutting the Internal, Two Lead Thread

WORKSHEET 1. Cutting the External, Two Lead Thread

Materials

4 in. of 1 in. dia. HRMS round

Procedure

Make a two-start, double lead unified screw thread having a pitch of .100 in. on a 1 in. dia. rod.

1. Cut off about 4 in. of the round stock, face and center drill one end.
2. Chuck about $\frac{1}{2}$ in. and place the dead center in the other end.
3. Take a skim cut to remove the scale. Take a micrometer measurement and record it. This will be the basis for your calculations for threads. <u>Note</u>: If CRMS is used, this step is not necessary.

4. Make an undercut about 3 in. from the tailstock end to terminate the thread. The depth is based on .100 in pitch.
5. Since the lead on a two-start thread is twice the pitch, it will be .200 in or 5 TPI. Set the quick-change gearbox to 5 TPI.
6. The two starts must be 180 degrees opposed, so you can use any of the several methods explained in your textbook. The most convenient, if your lathe is so designed, is to use the threading dial to make the division, using alternating positions on the dial. This should be tested with a scratch cut. This system will only produce two leads. Setting the compound parallel to the ways is a method in which there is no limit to the number of starts. However, the gibs must be tightened when you use that method.
7. Set up the lathe tool for threading and zero the micrometer collars. Take a scratch cut.
8. Advance the compound .100 in. (assuming you are using that method) and take a second scratch cut. All of the scratch marks should be equally spaced. The pitch of .100 in. (or 10 TPI on the gage) should be checked with a screw pitch gage.
9. Calculate the depth of the external thread in terms of the pitch, not the TPI set on the gearbox. The depth of cut on the cross slide is

$$P \times .613 = .100 \times .613 = .0613$$

or, if you are feeding in with the compound set at 29 degrees, the depth of cut is

$$P \times .708 = .100 \times .708 = .0708$$

10. Feed in, taking about .005 in. per pass on one lead at a time, leaving about .010 to .015 in. for finishing.
11. Now alternately, finish both leads to depth. Note: If you are using the compound at 29 degrees for infeed, be sure to back it out and set it to zero before you begin the second cut.

WORKSHEET 2. Cutting the Internal, Two Lead Thread

Materials

1 inch length of $1\frac{1}{2}$ in. dia. round or hexagonal stock

Procedure

Make one two-start nut to fit the just completed external two-start thread.

1. Set up the nut blank in a three-jaw chuck and face both sides. Chamfer the edges with a lathe tool.
2. Calculate the minor diameter. The formula is

$$d = D - (P \times .541 \times 2)$$

as explained in Project 10, "Internal Thread." If the major diameter of the external thread is slightly under one inch, because of the clean-up cut, use that figure in this formula for D.
3. Drill through the nut blank just under the calculated minor diameter. Bore to size.
4. Set up for internal threading and take the two scratch cuts as you did for the internal thread.
5. Cut each lead separately as you did with the external thread, leaving .010 in. for finish.

6. When you are near finish depth, make cuts alternately from one lead to the other, checking the fit with the external thread between each pass. When the thread will turn in all the way in both positions (180 degrees opposite the start), the job has been completed.
7. Clean up both pieces in solvent.
8. Turn in the grading sheet and finished project for your instructor's evaluation.

MACHINE TOOL PRACTICES

Name _____ Date _____

PROJECT 12. ACME THREADS

<u>Project Evaluation</u> (To be filled out by the instructor):

	Grade	
	Letter	Percent
1. Follows drawing (dimensions and tolerances)	_____	_____
2. Machining finishes	_____	_____
3. Mechanism or tool operates satisfactorily	_____	_____
4. General workmanship	_____	_____
Total grade	_____	_____

<u>Comments</u>:

Signed: _____
(Instructor)

MACHINE TOOL PRACTICES 85

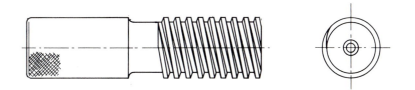

PROJECT 12. ACME THREAD

Objectives

1. Learn to cut external Acme threads on a lathe.
2. Learn to cut internal Acme threads on a lathe.

Outline for Study

Prior to starting each procedure for the project, study and complete Post-Tests for:

1. Making the external Acme thread: Section I, Units 14 and 16.
2. Making the internal Acme thread: Refer to Section I, Units 12 and 15.

Procedures

Begin the procedures for this project by completing the following worksheets.

Worksheet 1. External Acme Threads
Worksheet 2. Internal Acme Threads

WORKSHEET 1. External Acme Threads

Material

$1\frac{1}{2}$ in. piece of round HR steel $5\frac{1}{8}$ in. long

Procedure

Make a thread plug gage (Figure 1) for a $1\frac{3}{8}$-4 Acme thread.

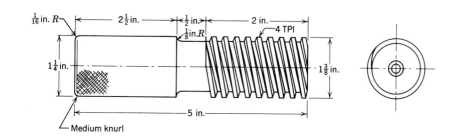

Figure 1.

1. Machine the plug gage blank as shown in Figure 1. Make the undercut 0.005 in. under thread depth.
2. Grind an Acme tool to the correct angles and dimensions as shown in your textbook.
3. Set up the lathe for cutting threads.
4. Set up and align the tool to the workpiece.
5. Cut the Acme thread to the correct depth as explained in the textbook.
6. Show the finished plug gage to your instructor before continuing on. Save this gage for checking the Acme nut in Worksheet 2.

WORKSHEET 2. Internal Acme Threads

<u>Material</u>

Piece of hexagonal or round 2 in. diameter CF steel $1\frac{1}{4}$ in. long

<u>Procedure</u>

Make an internal $1\frac{3}{8}$-4 Acme thread in a nut to fit the Acme plug gage.

1. Face the nut blank to length and drill a hole slightly less than the minor diameter.
2. Determine the minor diameter and record it here: _____. Bore the hole to the minor diameter of the thread.
3. Set up the Acme tool in the boring bar and align it with a gage.
4. Make a scratch cut and check the pitch.
5. Continue threading until the tool has reached the major diameter.
6. Clean out the threads and check the fit with the plug gage.
7. Continue to take cuts of one or two thousandths of an inch until the required fit has been made.
8. Chamfer the nut.
9. Turn in the grading sheet and finished project to your instructor for evaluation.

MACHINE TOOL PRACTICES

Name _____ Date _____

PROJECT 13. CARBIDE TOOL EXERCISES

<u>Project Evaluation</u> (To be filled out by the instructor):

	Grade	
	Letter	Percent
1. Follows drawing (dimensions and tolerances)	_____	_____
2. Machining finishes	_____	_____
3. Mechanism or tool operates satisfactorily	_____	_____
4. General workmanship	_____	_____
Total grade	_____	_____

<u>Comments</u>:

Signed: _____
(Instructor)

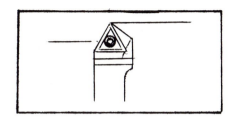

PROJECT 13. CARBIDE TOOL EXERCISES

Objectives

1. Learn to select carbide inserts and tooling for specific job requirements.
2. Use carbide tools on a lathe with various speeds, feeds, chipbreakers, and depths of cut to be able to discern machinability.
3. Be able to form chips for better chip control.

Outline for Study

Prior to starting the procedure for this project, study and complete Post-Tests for:

1. Using carbide tools: Section F, Unit 4.

Procedures

Begin the procedures for this project by completing the following worksheet.

Worksheet 1. Machinability Exercises
    Exercise 1. Feed and Chip Control
    Exercise 2. Speed and Its Influence on Cutting
    Exercise 3. Use of Formed Groove Chip Breaker

WORKSHEET 1. Machinability Exercises

Material

A 12 to 15 inch swing lathe and tooling
One piece of HR steel round bar $2\frac{1}{2}$ to $3\frac{1}{2}$ in. diameter approximately 14 in. long

Procedure

Complete the exercises that show the influences of feed, speed, depth of cut, and chip breaker geometry on chip control. Record your results.

EXERCISE 1. Feed and Chip Control

1. Mount the steel bar in a three- or four-jaw chuck. Support the tailstock end with a live center.

2. Select a steel cutting grade carbide tool having a separate adjustable chip breaker. Set a chip breaker width of $\frac{1}{16}$ in. List your carbide identification and selection. _____
3. Mount the toolholder securely in a quick-change tool post with minimum overhang. The cutting edge must be on center.
4. Select a cutting speed of approximately 300 to 350 FPM. Determine the RPM and record. _____
5. Set a feed of 0.002 in. per revolution.
6. Set the tool for a 0.060 in. depth of cut.
7. Make a cut $\frac{1}{2}$ to 1 in. long. Disengage the feed.
8. Record your findings on your Machinability Exercise Record Sheet. The chip will not be controlled, but should form a continuous ribbon.
9. Increase the feed to 0.004 in. per revolution; make a cut no more than 1 in. long. Record your findings and repeat this feed increasing exercise until the chip is broken into 9 or C shaped short lengths.
10. Continue to increase the feed until the chip becomes too tight. This condition is shown as a linking and welding of the chip, or a very short and tight letter C configuration. More horsepower is required to produce these chips.
11. Change chip breakers to a width of $\frac{3}{32}$ in.
12. Starting with the same depth of cut and speed as previously used and a feed of 0.010 in., take cuts as before and record the results after each cut. With the wider chip breaker, the chips should be controlled at heavier feeds.

EXERCISE 2. Speed and Its Influence on Cutting

1. Set up the lathe as for the first exercise.
2. The tool should have a $\frac{1}{16}$ in. wide chip breaker.
3. Set a feed that gave good chip control in the first exercise and a 0.060 in. depth of cut.
4. Set a speed of 100 FPM and make a 1 in. long cut. Record the results; fill in all spaces on the record sheet.
5. Make successive cuts, increasing the speed by 50 FPM for each cut to 400 FPM, then continue with 100 FPM increments. Record your findings after each cut.
6. How did the speed changes affect the surface finish and chip control?

EXERCISE 3. Use of Formed Groove Chip Breaker

1. Set up the lathe as for prior exercises.
2. Select a tool with a grooved chip breaker.
3. Set a 0.060 in. depth of cut.
4. Set a speed that gave good results in the previous exercise.
5. Start with a feed of 0.004 in. and make a cut. Record the results.
6. Increase the feed with 0.002 in. increments, taking cuts and recording your findings. Stop making feed increases when the chips are getting too tight.
7. How effective are formed groove chip breakers? Why?
8. Make a chart as shown on page 91 on a full size sheet of paper and record your results.
9. Turn in the grading sheet and the record sheet you have prepared to your instructor for evaluation.

Machinability Exercise Record Sheet

Work Material					Tool Identification		
FPM	RPM	Feed	Depth of cut	Chip breaker width	Chip form	Surface finish of work	Tool wear

MACHINE TOOL PRACTICES

Name _____ Date _____

PROJECT 14. HYDRAULIC JACK

Project Evaluation (To be filled out by the instructor):

	Grade	
	Letter	Percent
1. Follows drawing (dimensions and tolerances)	_____	_____
2. Machining finishes	_____	_____
3. Mechanism or tool operates satisfactorily	_____	_____
4. General workmanship	_____	_____
Total grade	_____	_____

Comments:

Signed: _____
(Instructor)

MACHINE TOOL PRACTICES   95

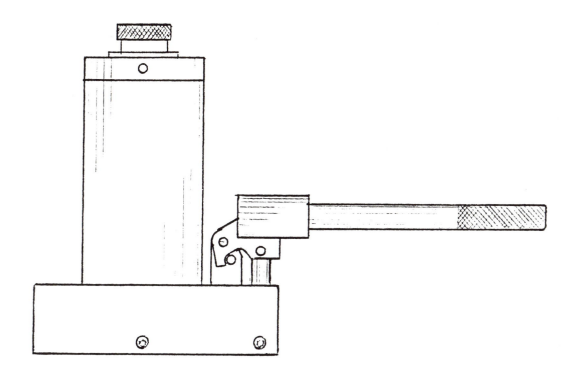

PROJECT 14. HYDRAULIC JACK

Objectives

1. Learn to square a rectangular block in a horizontal mill or by facing in a lathe.
2. Bore, counterbore, and internally thread in a blind hole.
3. Drill, ream, and counterbore; also learn deep drilling techniques and drilling a hole on an angular surface.
4. Internal and external threading.
5. Using a vertical mill to make a slot and a horizontal mill to make a notch.
6. Using layout, machining, and welding techniques to produce a useful device.

Outline for Study

Prior to starting each procedure for this project, study and complete Post-Tests for:

1. Making the base: Section J, Units 1, 2, 3, and 4; Section K, Units 1 through 7; also refer to Section E, Units 1 and 2; Section H, Units 5, 6, and 7; and Section I, Units 10, 11, and 13.
2. Making the cylinder tube: Refer to Section I, Unit 12.
3. Making the piston and assembly: Refer to Section I, Unit 16.
4. Making the pump plunger: Refer to Section J, Unit 3.

Procedures

Begin the procedures for this project by completing the following worksheets.

Worksheet 1. Making the Base   Worksheet 4. Making the Reservoir Tube
Worksheet 2. Making the Cylinder Tube   Worksheet 5. Making Pump Pivot Support
Worksheet 3. Making the Piston Rod Assembly Worksheet 6. Making pump Handle Socket

Worksheet 7. Making the Pump Plunger   Worksheet 9. Making the Cap
Worksheet 8. Making the Needle Valve   Worksheet 10. Assembly of the Jack

Many of the major parts of this project can be made of aluminum alloy for weight reduction. However, all of the parts can be made of steel with one or two exceptions that are noted on the drawings. No castings are necessary since all parts are machined from standard stock material.

WORKSHEET 1. Making the Base

Material

$7\frac{1}{2}$ inches long, 2 x 4 in. mild steel or aluminum alloy bar

Procedure

Ask your instructor which material you should use, mild steel or aluminum alloy.

1. Cut off $7\frac{1}{2}$ in. of 2 x 4 in. bar stock.
2. Machine the ends square in a lathe or on a horizontal mill and to 7 in. in length. Face or mill the other sides slightly undersize if necessary to clean up the scale.
3. Lay out the top surface and center punch where drilling or boring will take place.
4. Set up the base in a four-jaw chuck for making the bore, counterbore, and threads. Set up to your layout punch mark.
5. Rough drill to about $1\frac{1}{2}$ in. diameter, being careful not to go too deep.
6. Set up for boring with an angled tool so that a flat bottom can be produced. Bore to the <u>minor</u> diameter of the 20 TPI. Remember, this will be 1.750 in. <u>minus the double depth of the thread</u>.
7. Make the thread relief in the bottom of the bore. This will require a specially made tool bit.
8. Make the two counterbores to within tolerance dimensions (Figure 1).
9. Cut the thread to the correct depth (Figure 2).
10. Remove the base from the lathe chuck.
11. Apply layout dye to the left and front sides of the block (when facing the valve and pump end). Scribe center lines on the end and side for drilled holes and for the angles of internal drilled holes. See Section A-A of the base drawing on page 114. The 18 degree angled hole may vary slightly, depending on the accuracy of your setup. These external angular layout lines should help you to make a more accurate setup.
12. Lay out and center punch all of the drilling locations on the base.
13. Set up the base in a drill press top surface up. Make sure there is clearance for the drill to come through the bottom of the $\frac{33}{64}$ in. hole so it will clear the table or drill press vise.
14. Pilot drill both holes, taking care to set the limit stop so the $\frac{3}{16}$ in. hole will not go in too far. Set the limit stop and drill a $\frac{27}{64}$ in. hole $1\frac{7}{16}$ in. deep, as it is called off on the drawing and then ream with a $\frac{7}{16}$ in. machine reamer. Use cutting oil. Chamfer the top of the reamed hole; this will allow the pump plunger seal to be installed without cutting it. This reamed hole must have a smooth finish, so the reamer should be checked in a piece of scrap metal before using it in the base.
15. Drill the $\frac{33}{64}$ in. hole. Chamfer.
16. Turn the base over and counterbore the drilled hole to $1\frac{1}{4}$ in. dia., $\frac{5}{8}$ in. deep This counterbore provides space for a $\frac{1}{2}$ in. nut and room for a socket wrench.

Figure 1. Making the counterbore in the base (Lane Community College).

Figure 2. Cutting the internal thread in the base (Lane Community College).

Figure 3. Milling the slot on the base; note the correct direction of cut (Lane Community College).

17. The 45 degree notch for the needle valve should be milled at this time before preceeding further with the drilling (Figure 3).
18. Drill all of the oil porting holes as shown on the drawing. Be sure to drill the angular $\frac{1}{16}$ in. hole before drilling the $\frac{3}{16}$ in. hole it connects with, since the small diameter drill tends to grab and break off when it enters another hole at an angle; when that happens, it is extremely difficult to remove. The drill diameters and depth of drilling are both extremely important where the steel ball check valves are located for the successful operation of the jack. When drilling deep holes, be sure to remove the drill from time to time to remove the chips. This is called "pecking." If this is not done, the drill will jam and either break off or be very difficult to remove.
19. The $\frac{1}{8}$ in. 18 degree angular hole must be drilled accurately so as to intersect only with the $\frac{1}{4}$ in. hole. This hole returns oil to the reservoir when the needle valve is opened. An excellent way to drill this hole is to set up the base in a vise in a vertical milling machine as shown in Figure 4. A special extension tool is used to bring a $\frac{1}{8}$ in. end mill to the proper location. The end mill is used to make a starting hole for the $\frac{1}{8}$ in. long shank drill. The long shank drill is then placed in a drill chuck and the hole is drilled (Figure 5).

Figure 4. A special extension collet is needed to make the starting hole for drilling the 18 degree angular $\frac{1}{8}$ in. hole (Lane Community College).

Figure 5. The extension collet is exchanged for a chuck and an extra long $\frac{1}{8}$ in. drill to make the hole (Lane Community College).

20. Tap the $\frac{1}{2}$-24 thread; finish it with a bottoming tap. Tap all of the $\frac{1}{8}$ in. pipe threads to a depth sufficient to cause the pipe plugs to screw in far enough to cover all of the threads on the plug when it has been tightened. If socket type plugs are used, there should be no part of the plug projecting from the base.
21. Chamfer all edges of the base $\frac{1}{16}$ in. using a file. Draw filing works best for this. The base is now complete except for assembly of parts which will be done later.

WORKSHEET 2. Making the Cylinder Tube

Materials

$7\frac{1}{2}$ inches of $1\frac{1}{2}$ ID x $1\frac{7}{8}$ OD in. seamless steel tubing or equivalent

Procedure

1. Seamless tubing, $1\frac{1}{2}$ in. ID and $1\frac{7}{8}$ in. OD, with an acceptable inside finish can be purchased for this part. However, seamless tubing is very expensive. There are alternatives. Heavy wall water pipe will not work because it is butt-welded with a seam where it is likely to burst under high pressure, but solid hot or cold rolled shafting can be made into a cylinder by drilling almost to size from each end and then boring and reaming to get a smooth finish. It may be necessary to hone or polish the cylinder to get an acceptable finish. Drilling out a solid shaft is usually much less expensive than using seamless drawn tubing.
2. The ends are turned and threaded as shown in the drawing. The thread on the bottom end must fit the threads in the base. The No. 60 drilled hole in the top end is for relieving the pressure if the jack ram is pumped too far out; it acts as a safety valve to protect the jack from damage. A hole larger than a No. 60 could cause damage to the cup packing.
3. Chamfer the edges of the cylinder bore. This will allow the cup seal to enter without shear damage. The cylinder tube is now complete.

WORKSHEET 3. Making the Piston Rod Assembly

Materials

$8\frac{1}{2}$ in. of $1\frac{3}{8}$ in. CR or larger diameter shaft turned to size $\frac{1}{2}$-20 jam nut

$4\frac{1}{2}$ in. of $1\frac{1}{2}$ in. HRMS round $1\frac{1}{2}$ in. dia. bronze stock for guide

Procedure

The Extension Screw

1. It is always best, when both external and internal threads are to be made on the lathe with single point tools, to begin with the external thread. This provides an accurate test thread plug. External threads are more predictable than internal threads because of the spring of the boring bar.
2. Hot or cold rolled shaft, $1\frac{1}{2}$ in. dia., may be used for this project. It should be cut off 1 in. longer than finish size for chucking.
3. Center drill one end and knurl the other end if desired.

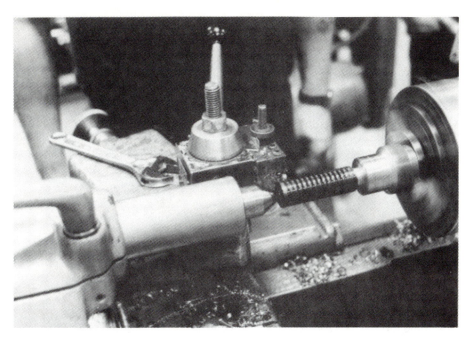

Figure 6. Cutting the Acme thread on the extension screw (Lane Community College).

4. Turn to thread size and make an undercut for thread relief.
5. Cut the standard Acme threads to depth and chamfer threads (Figure 6). Part off to length.
6. The cross-hatching on the top end may be done on a horizontal mill with a 90 degree angular cutter.
7. The extension screw is now completed.

The Piston Rod

8. The piston rod may be turned from $1\frac{1}{2}$ in. hot or cold rolled steel shaft finished by filing and polishing.
9. The rod must now be chucked in an accurate three-jaw chuck or a four-jaw chuck, using a dial indicator. Be sure to use jaw pads made of soft copper or aluminum to protect the finish.
10. Face the rod to a length and turn the steps for the piston and nut. Thread $\frac{1}{2}$-20 for a standard jam nut. The bronze piston guide and steel washer can be made with another setup at a later time.
11. Turn the rod end for end and drill under the minor diameter of the thread to the depth specified.
12. Set up a boring bar and bore the 5 TPI Acme thread minor diameter (or alternate). Now make the counterbore and thread relief as shown.
13. Make the internal Acme thread with the boring bar (Figure 7). When you have reached the correct depth, check the thread with the extension screw. You will probably need to make several "free cuts" in order to have a fit because of boring bar spring. Break all corners on the rod. The piston rod is now completed.

Figure 7. Making the internal Acme thread in the piston rod (Lane Community College).

WORKSHEET 4. Making the Reservoir Tube

Material

$6\frac{1}{4}$ inches of 3 in. standard aluminum or steel pipe

Procedure

1. Standard 3 in. aluminum or steel pipe can be used for this part. The inside diameter may be bored out to an approximate size (Figure 8). This inside dimension is not critical except for about $\frac{1}{4}$ in. in on the top end when it fits over the cap. The bottom end is critical only on the outside for a short distance (Figure 9).
2. The OD should be turned to size and finished by filing and polishing. Break all corners with a file. The reservoir tube is now completed.

WORKSHEET 5. Making the Pump Pivot Support

Material

$3\frac{1}{2}$ inches of $\frac{3}{4}$ x 1 in. rectangular cold rolled bar stock

Procedure

1. Cut off $3\frac{1}{2}$ inches of $\frac{3}{4}$ x 1 in. rectangular cold rolled bar stock. If none is available, it can be milled to size from 1 in. square bar.
2. Lay out for center drilling at the center of the cylindrical part, that is $\frac{3}{8}$ in. from one edge. Lay out for drilling on the side and lay out the radius (Figure 10).

Figure 8. Boring the ID in the reservoir tube (Lane Community College).

Figure 9. Turning the OD of the reservoir tube. This operation takes two setups using this method. A large expandable mandrel would work better (Lane Community College).

Figure 10. Laying out the radius on the pivot pump support (Lane Community College).

3. Drill the $\frac{5}{16}$ in. hole. Drill the $\frac{1}{4}$ in. hole first with a C drill and ream to $\frac{1}{4}$ in. This is done so the drill rod stop pin will be a drive fit so it will remain in place. Drill the center hole in the end.
4. Saw the radius and angular surface and finish by filing.
5. Set up the rectangular end in a four-jaw chuck in the lathe using protective pads. About 2 in. should be extended from the chuck. Insert the dead center and set up the work parallel to the lathe by fastening a dial indicator to the carriage and moving the indicator along the length of the work. When the indicator has no movement in both axes, the work is correctly set up. Turn the $\frac{1}{2}$ in. diameter and cut the $\frac{1}{2}$-13 threads. Remove the work, break all corners, and install the $\frac{1}{4}$ in. pin. If the pin is loose, lightly knurl it in its center area to enlarge it and then drive it in place.
6. Three CR (or drill rod) steel pins will be needed. One $\frac{1}{4}$ in. diameter, $1\frac{1}{4}$ in. long; the second one $\frac{5}{16}$ in. in diameter, $1\frac{1}{2}$ in. long; and the third one should be $\frac{1}{4}$ x $1\frac{1}{2}$ in. with cotter pin holes; it is for the pump plunger. Both $1\frac{1}{2}$ in. long pins will need a $\frac{3}{32}$ in. hole near each end for cotter pins.

WORKSHEET 6. Making the Pump Handle Socket

<u>Materials</u>

7 inches of 1 x $\frac{3}{4}$ in. or 1 x 1 in. keystock; 10 inches of $\frac{3}{4}$ in. dia. CR round;
2 inches of $1\frac{1}{4}$ in. dia. CR shaft; $2\frac{1}{2}$ inches of $\frac{1}{4}$ x 2 in. HR flat bar (two pieces)

Procedure

1. The side plates are made of $\frac{1}{4}$ in. hot or cold rolled plate or flat bar stock. It works out well to use $\frac{1}{4}$ by 2 in. HR bar stock. Cut two pieces about 3 inches long and fasten one on top of the other by making short weld beads on the edges. These welds will later be part of the scrap when the part is sawed out. An alternate method is to use longer pieces and clamp them together with C-clamps.
2. Lay out the hole locations and center punch. Lay out cutting lines.
3. Drill the $\frac{5}{16}$ and the $\frac{1}{4}$ in. holes.
4. Using a vertical bandsaw, saw the layout shape of the side plates. The last cut should separate the two plates from the last weld so that you now have two identical plates. Finish the edges with a file and chamfer all sharp edges.
5. Cut off 2 in. of $1\frac{1}{4}$ in. CR round stock. Set up in a lathe in a three-jaw chuck.
6. Face both ends and chamfer corners. Drill $\frac{46}{64}$ in. to depth. Remove from the lathe.
7. Set up the side plates for welding by clamping a $\frac{3}{4}$ in square bar and a shim of 0.020 to 0.030 in. between the plates. Also, $\frac{5}{16}$ in. and $\frac{1}{4}$ in. pins or bolts should be inserted in the holes to assure alignment. With the two plates clamped together in this fashion, place the socket in the offset position as shown in the drawing. Tack weld it in place. Weld the plates to the socket with a good penetrating rod or wire and fill out the weld enough that it can be finished by filing. The pump handle can now be made according to the specifications on the drawing. The pump handle and socket assembly are now complete.

WORKSHEET 7. Making the Pump Plunger

Material

$2\frac{1}{2}$ inches of $\frac{3}{4}$ x 1 in. HRMS bar

Procedure

1. Cut off about $2\frac{1}{2}$ in. of $\frac{3}{4}$ x 1 in. hot or cold rolled rectangular stock.
2. The milling operation should be done before turning the plunger and because the piece can be gripped better in a vise. Set up in a vertical mill with a $\frac{1}{4}$ in. two flute end mill. Make the slot by milling $\frac{1}{8}$ in. deep for each pass until the slot is through the piece. If the end mill is too short, cut one-half depth and turn the part over in the vise and cut to center to meet the opposite cut. About 0.005 in. should be skimmed off one side so the $\frac{1}{4}$ in. pin will slide freely in the slot.
3. Remove the work from the mill and set it up in a four-jaw chuck in a lathe. Center drill with a No. 1 center drill and set up the dead center. Turn the plunger to size so it will have a sliding fit in the $\frac{7}{16}$ in. reamed hole in the base. It should have a smooth finish. Now turn the $\frac{3}{16}$ in. diameter on which the plunger seal fits. This dimension may have to be adjusted to the particualr plunger seal available to you. The thread on the end is 10-32 and a standard nut is used on it to hold the seal in place.
4. Remove the plunger from the lathe and file finish the external radius. Break all sharp corners. The pump plunger is now complete.

WORKSHEET 8. Making the Needle Valve

Material

1 inch of $\frac{5}{8}$ in. hexagonal bronze stock;   3 inches of $\frac{3}{8}$ in. dia. CR round bar

Procedure

1. The valve nut should be made of   in. bronze hexagonal stock. If none is available, use two or three inches of round stock and mill six flats where needed.
2. Set up the stock in a lathe, extending about 1 in. from the chuck.
3. Drill the $\frac{1}{8}$ in. hole $\frac{3}{4}$ in. deep.
4. The O-ring groove can be made by using a specially made boring bar. The bar can be made of $\frac{3}{16}$ or $\frac{1}{4}$ in. drill rod, turned to a slight taper on one end. It is then heated to red heat with a torch on the small end and bent about $\frac{1}{8}$ in. 90 degrees. It is then shaped with a grinder to a grooving tool shape that will fit into the $\frac{3}{16}$ in. hole. The tool is hardened and tempered to a light straw color. The O-ring groove should be made about 0.055 to 0.060 in. deep and about 0.080 to 0.090 in. wide.
5. Turn the OD to $\frac{1}{2}$ in. and make an undercut for the threads. Make the $\frac{1}{2}$-20 thread with a single point tool (not with a die). Check the thread with a nut.
6. Now part off the valve nut slightly longer than finish length.
7. Turn the nut around and set it up in a chuck, using protective material around the finished threads. A standard nut, split on one side with a hacksaw, is one way to protect threads from chuck jaws. The $\frac{3}{16}$ in. drilled hole must have no more than 0.002 in. indicator runout when set up.
8. Face off the nut to length and chamfer.
9. Using the ¼-28 tap drill, make the hole $\frac{1}{2}$ in. deep for tapping.
10. Hold the ¼-28 tap in a chuck in the tailstock. Leave the tailstock free to slide on the ways and start the tap, turning the chuck on the headstock spindle by hand. Do not use power. When the tap is near the bottom of the tap drilled hole, remove it and substitute a bottoming tap. Complete the threads to the correct depth. This completes the needle valve nut.
11. The needle valve is turned from CR mild steel. Do not use hot rolled steel or spheroidized drill rod as they are too soft and will break when twisting pressure is applied.
12. Chuck the $\frac{3}{8}$ in. stock with about $2\frac{1}{4}$ in. extending from the chuck. Turn the diameters as shown on the drawing and make the taper with the compound method.
13. Part off the valve screw, reverse it in the collet or chuck, gripping the $\frac{3}{8}$ in. diameter, and chamfer as shown on the drawing.
14. Drill the cross hole so a $\frac{1}{8}$ in. pin will drive in. If the pin is loose, knurl the center part of the pin and then drive it in place. The needle valve is now complete.

WORKSHEET 9. Making the Cap

Material

$1\frac{1}{2}$ inches of $3\frac{1}{2}$ in. dia. HRMS round

Procedure

1. The material for the cap can be HR steel or aluminum alloy, as desired. Cut it off $\frac{1}{4}$ in. longer than the finish length.
2. The end with the $\frac{1}{2}$ in. setp and $1\frac{3}{4}$ in. dia. threads should be turned first. The $3\frac{1}{8}$ in. step, the bore, and the threads should all be completed in one setting before reversing the work. This is to assure alignment of the working parts.
3. Chuck about $\frac{1}{4}$ in. of the piece in a four-jaw chuck (for better holding power) and drill through with a size under the $1\frac{7}{16}$ in. dia. Face off the end to true the cut surface.
4. Set up a boring bar and bore to $1\frac{7}{16}$ in. dia. Now, bore to a shoulder $\frac{3}{4}$ in. deep and to the <u>minor</u> diameter of the thread, not to the dimension shown on the drawing. Make a thread relief undercut at the bottom of the threads.
5. Set up for threading. You will probably need to make a special threading tool bit for an angular boring bar holder. Make the thread, checking it frequently with the <u>top</u> thread of the cylinder tube.
6. Turn the steps as shown on the drawing (Figure 11). Turn the OD to as near the chuck jaws as possible without risking any possibility of running into the turning jaws with the tool (Figure 12).

Figure 11. Student checking the fit on the cap with the reservoir tube. An experienced hand need not make this kind of check but would depend on precision measuring and machining procedures (Lane Community College).

Figure 12. Turning the OD on the cap (Lane Community College).

7. Remove the workpiece, turn it end for end, and rechuck it, using protective material (jaw pads). True the work both radially and axially with a dial indicator with no more than 0.002 in. indicator runout.
8. Face off the work to length and finish turn the OD to match the previously turned position. Make a large chamfer on the outer edge.
9. Bore the $1\frac{3}{4}$ in. packing gland dia. and the <u>minor</u> dia. of the thread $\frac{5}{16}$ in. deep. No undercut should be used. Remove the tool at the end of each pass. Cut the thread to depth. Remove the workpiece.
10. A bronze nut may be turned from bronze tubing stock or solid stock if tubing is not available. Set up in a three-jaw chuck.
11. All of the boring, turning, and threading operations should be done and the thread checked for fit in the cap.
12. Remove the workpiece and set up the bronze stock in a vertical position in a horizontal or vertical mill. A three-jaw chuck horizontally mounted on a plate is a useful holding device for this operation. Set up a ¼ in. wide cutter and mill the slots as shown on the drawing.
13. Return the workpiece to the lathe and rechuck it for cutoff operations. Set up a parting tool and cut to half depth. Now, using a file, chamfer the edge being cut off. Continue parting off. Deburr the inside diameter on each end.
14. Drill the angular overflow hole in the cap as shown. The cap assembly is now complete.

WORKSHEET 10. Assembly of the Jack

Materials

Assorted springs, pins, bearing balls, and seals as called for on the drawing and as noted in the instructions.

Procedure

1. All parts should be cleaned in solvent and holes and passageways blown out with air. Wear safety glasses when you do this.
2. Seat the $\frac{1}{4}$ in. and $\frac{5}{16}$ in. steel ball check valves. First set the jack base on edge on a solid surface and drop the $\frac{1}{4}$ in. ball in place. Using a soft steel rod as a punch, give the ball several light blows with a hammer. Now do the same with the $\frac{5}{16}$ in. ball.
3. Prepare a very light spring $\frac{3}{4}$ in. long (a ball point pen spring is about right) and place it between the $\frac{1}{4}$ in. and $\frac{5}{16}$ in. ball. If this spring is not light enough, the pump will not prime or operate correctly. Now place an $1\frac{1}{4}$ in. long spring over the $\frac{5}{16}$ in. ball (the spring pressure is not so critical here, almost any spring will do) and screw in the pipe plug.
4. Put all plugs in place. A sealing compound on the threads will help to eliminate leaks.
5. Insert the $\frac{3}{16} \times \frac{5}{16}$ in. O-ring in the needle valve nut. This can be accomplished with two flat end punches or rods. Screw the needle valve nut into the base. Screw the needle valve in to where it seats. If it does not seat, the jack will not hold in place under pressure. This can be checked by putting prussian blue or layout dye on the tapered end of the needle valve and turning it in until it is tight. Remove the needle valve screw and if the dye has been removed in a circle all around the taper, the valve is seating.
6. Assemble the piston rod, guide, cup seal, washer, and nut. Tighten the nut firmly, but not so tight that the seal is extruded (squeezed out). Using a center punch, stake (distort the threads) the $\frac{1}{2}$ in. dia. threaded end in four places to prevent the nut from coming off. If the cup seal is leather, it should be soaked in oil for an hour or so before assembly; if it is the neoprene-fabric type, no soaking is needed.
7. Oil the inside of the cylinder tube and put oil on the rod and piston assembly. Insert the top end of the rod into the bottom end of the tube. Be sure there is adequate chamfer in the tube; the cup seal will be damaged if it is not chamfered sufficiently. Now tap the rod and seal into the tube, using a soft hammer.
8. Oil the threads on the bottom end of the cylinder tube and screw it into the base. A sealing compound should be placed on the bottom face of the tube to prevent internal leakage.
9. Place the reservoir tube in the recess. A sealing compound or gasket should also be used here to prevent leakage.
10. The cap can now be screwed in place. Oil the threads and use sealing compound on the end of the reservoir tube. Tighten the cap with a hook spanner.
11. Place one ring of $\frac{3}{16}$ in. square rope packing in the packing gland and tighten the gland nut over it.

12. Screw the extension screw into the piston rod end.
13. Assemble the pump pivot support into the base using a $\frac{1}{2}$-13 nut. Make it finger tight until other parts are in place.
14. Assemble the pump plunger with its seal and nut. Stake the nut.
15. Make sure there is sufficient chamfer in the $\frac{7}{16}$ in. hole. Oil the pump seal and the hole, then insert the seal carefully so it will not be damaged. Make sure it will slide to the bottom freely.
16. Assemble the pump handle socket in place with the $\frac{1}{4}$ x $1\frac{1}{2}$ in. and the $\frac{5}{16}$ x $1\frac{1}{4}$ in. pins. These pivots should be oiled as well as the slot in the pump plunger.
17. The jack is now ready to be tested. Unscrew the filler plug on the side and lay the jack on its side, filler hole up. Put about one pint of automatic transmission oil or a good grade of 20W lube oil in the reservoir. Replace the filler plug.
18. Place the jack handle in the socket and pump several strokes. If the pump does not work, remove the pump plunger and fill the hole with oil and replace the plunger. This should prime the pump.
19. If all the holes are drilled correctly and the parts made according to the drawing dimensions, the jack will work very well. There are only two or three reasons for the jack to be inoperable:

    (a) The intake (suction) ball valve spring is too strong or too long. There is only atmospheric pressure to move the oil to the pump through this ball check. The cure is to shorten the spring or get a softer one.
    (b) The needle valve is not seated. This could be caused from drilling the $\frac{1}{4}$ in. hole too deep. The only cure for this problem would be to make a special needle valve with a longer extension than the one on the drawing.
    (c) The balls are not properly seated and they leak back when pressure is applied. If this happens, remove the balls and springs and reseat the balls with a soft punch and hammer.

20. Your 5-ton hydraulic jack has now been completed. These jacks have been tested under load, and will actually raise 7 tons. It is advisable not to use a jack handle longer than the one shown in the drawing as this may damage the jack.
21. Ask your instructor to help you test your completed hydraulic jack.
22. Turn in the grading sheet and finished hydraulic jack to your instructor for evaluation.

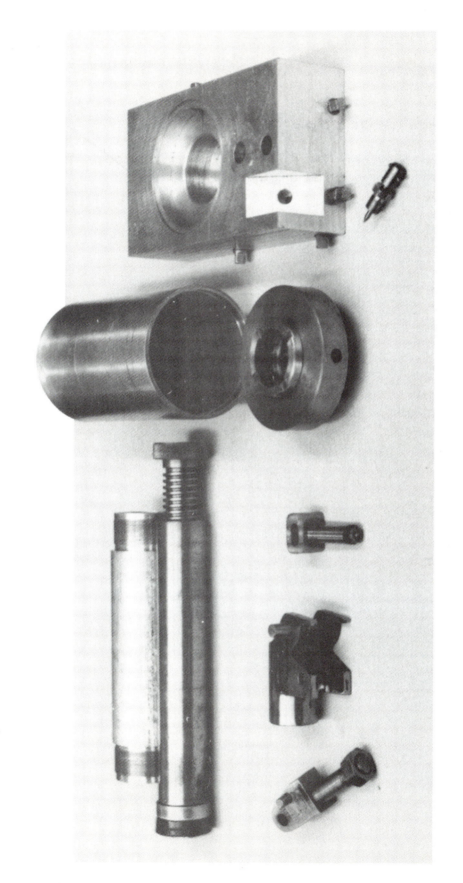

Figure 13. The parts of a hydraulic jack.

111

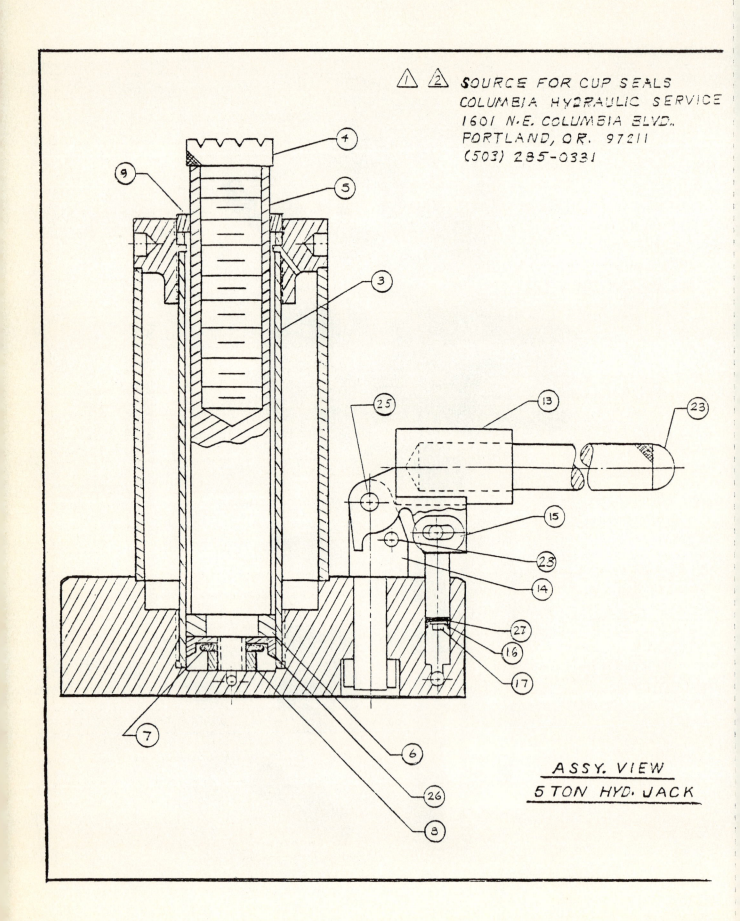

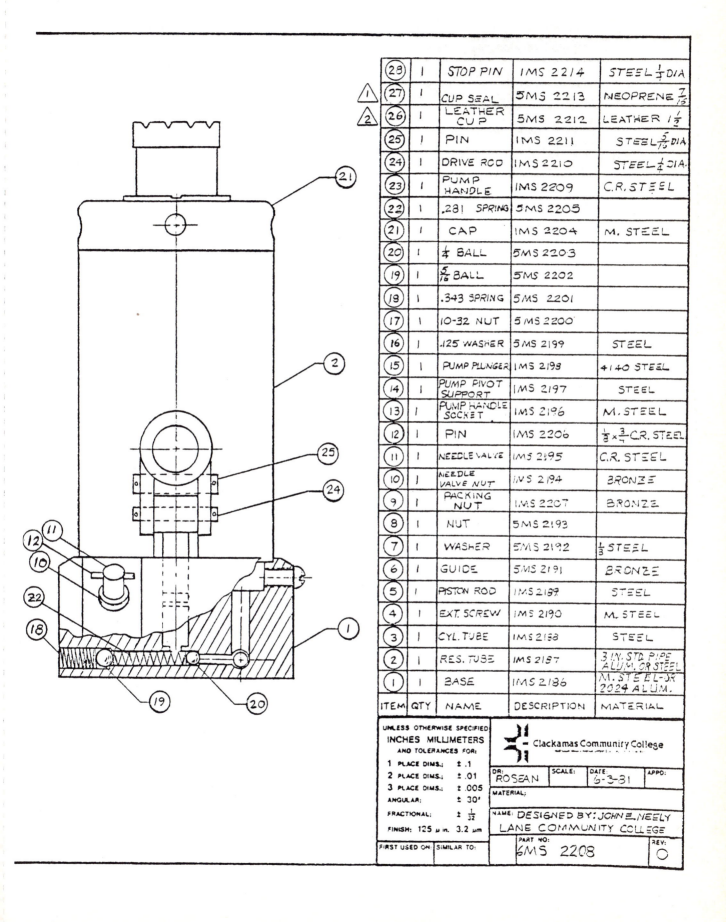

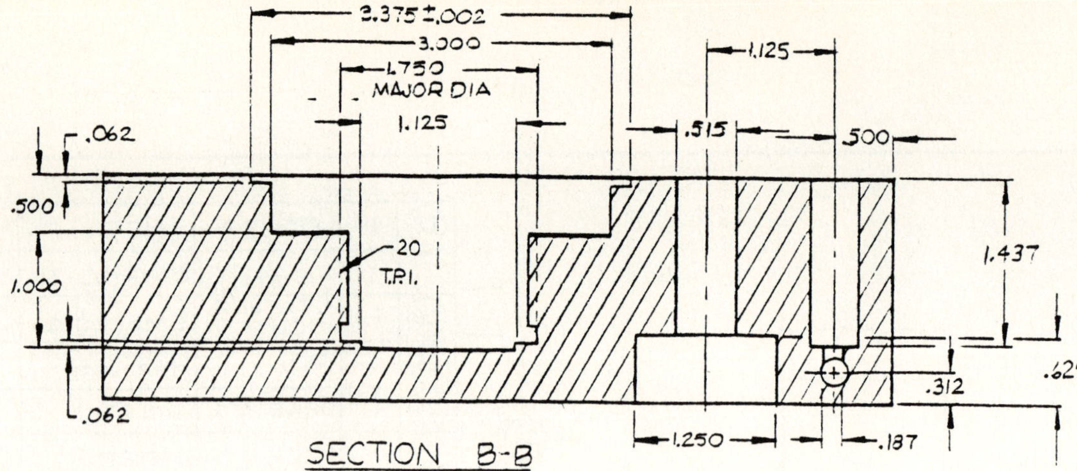

SECTION B-B

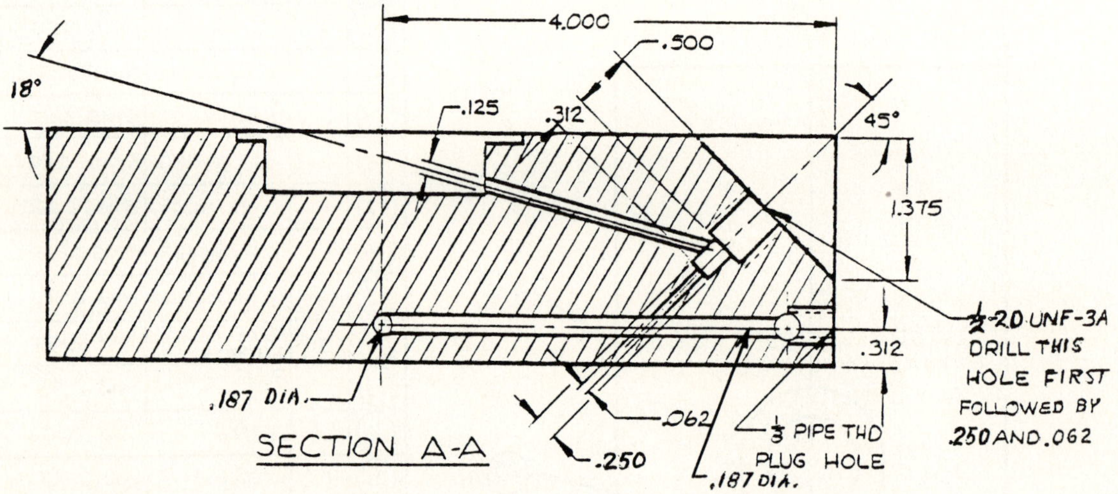

SECTION A-A

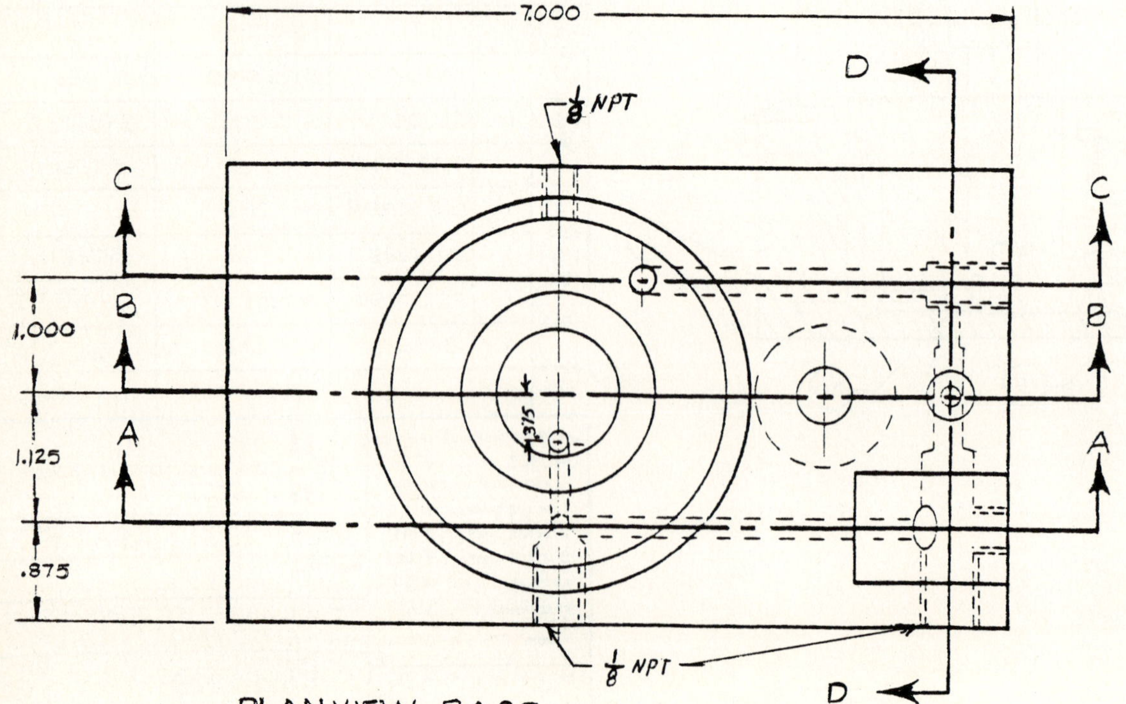

PLANVIEW- BASE

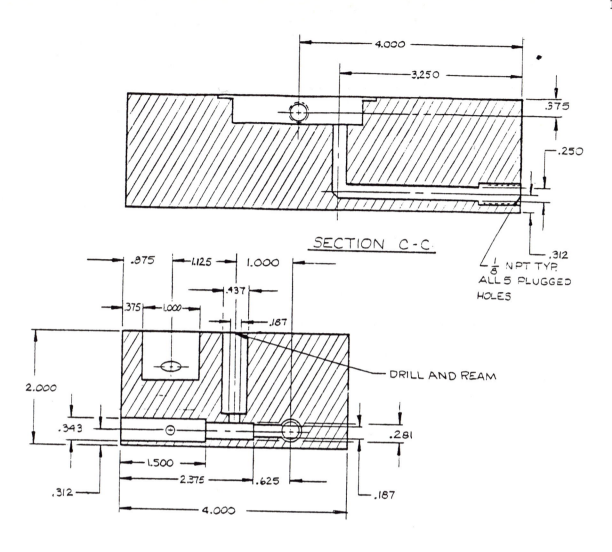

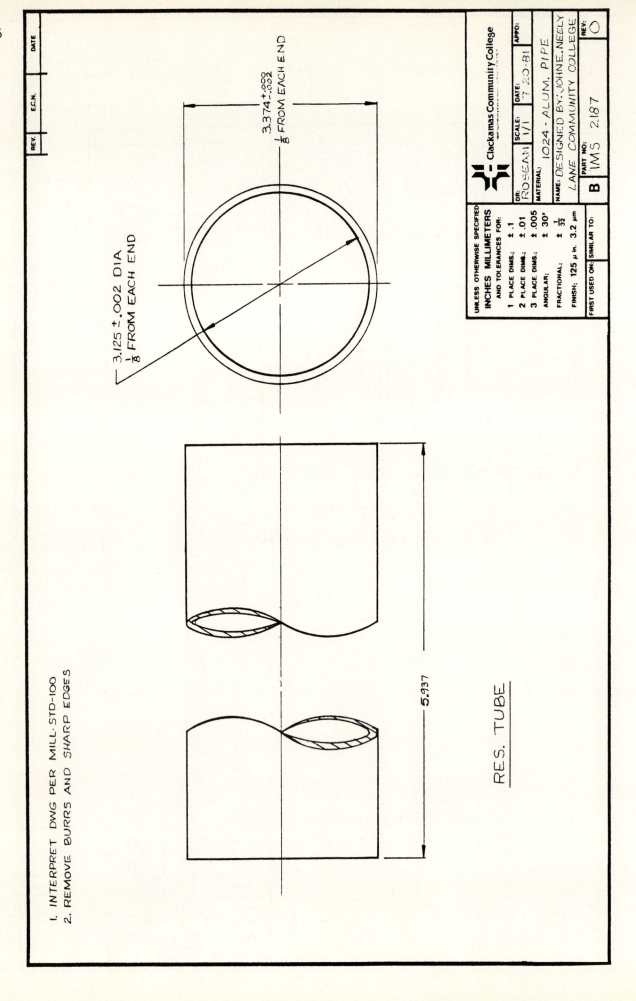

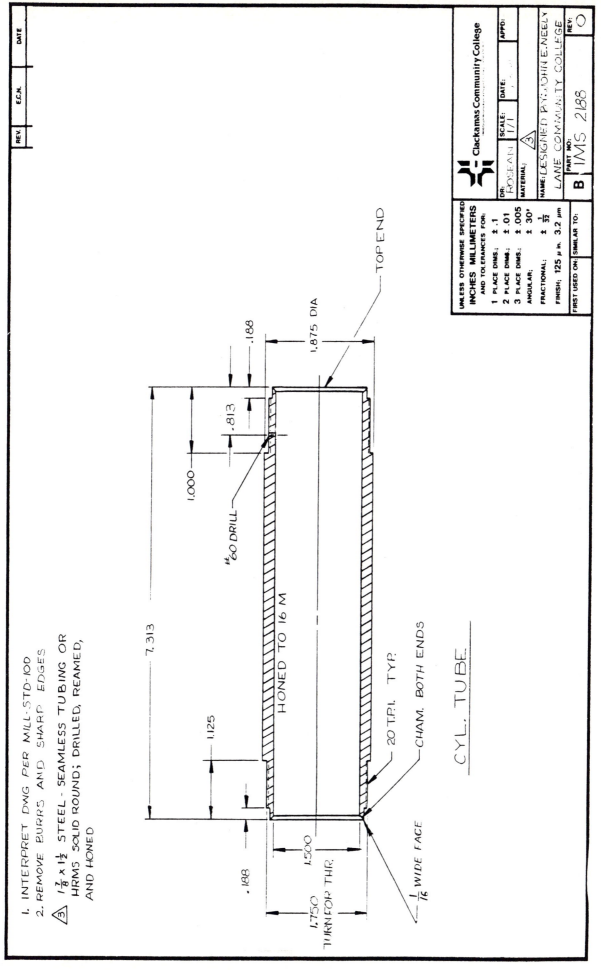

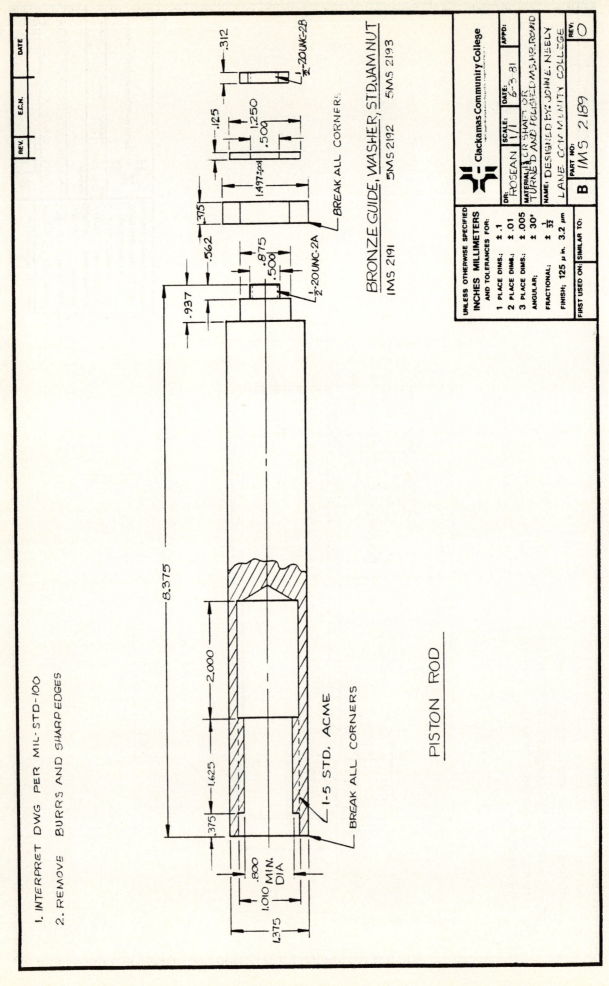

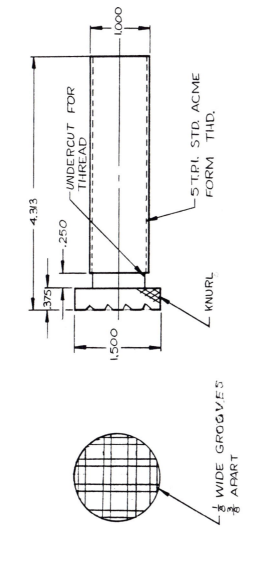

EXT. SCREW

1. INTERPRET DWG PER MILL-STD-100
2. REMOVE BURRS AND SHARP EDGES

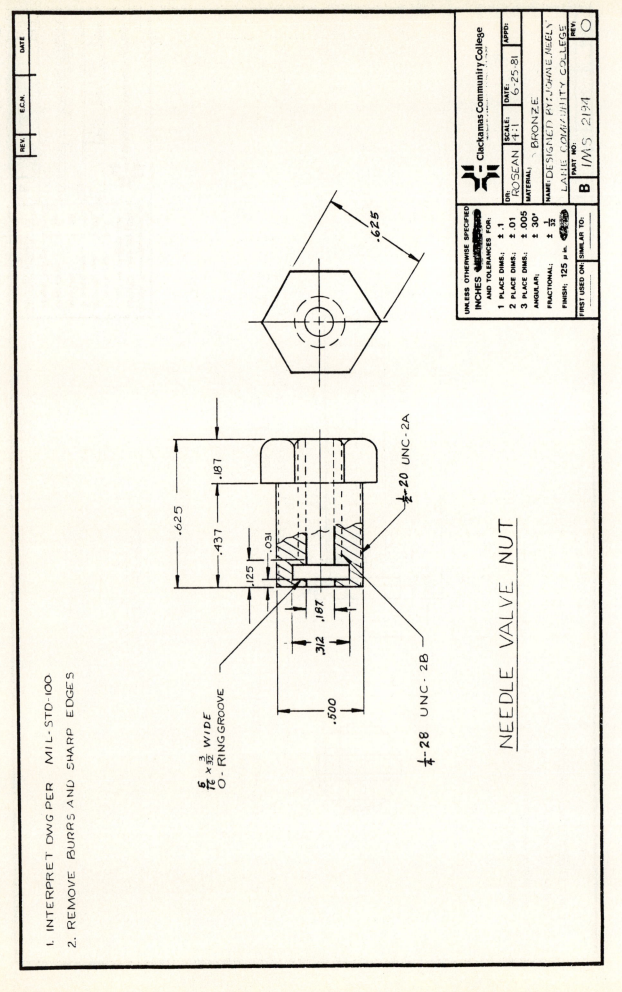

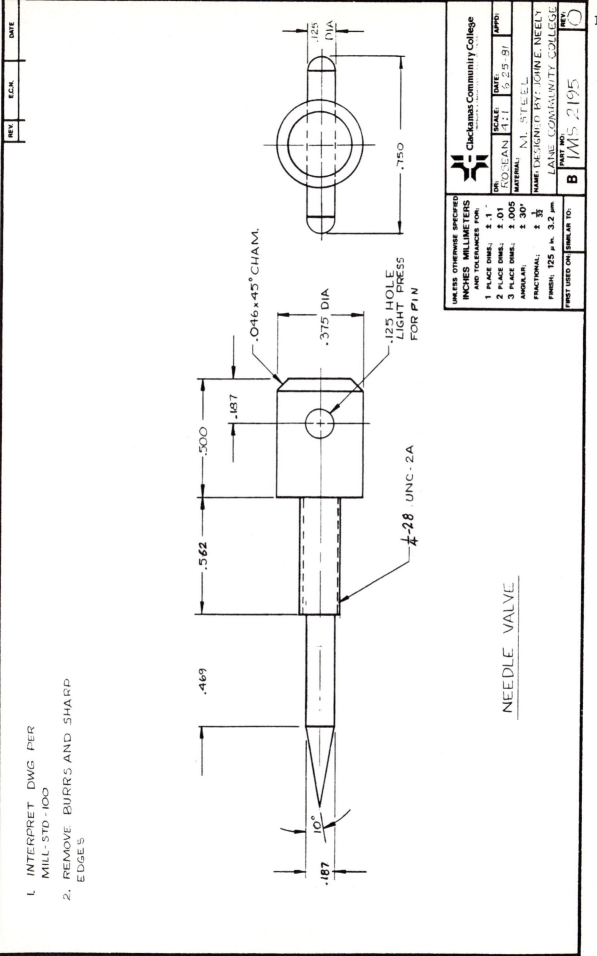

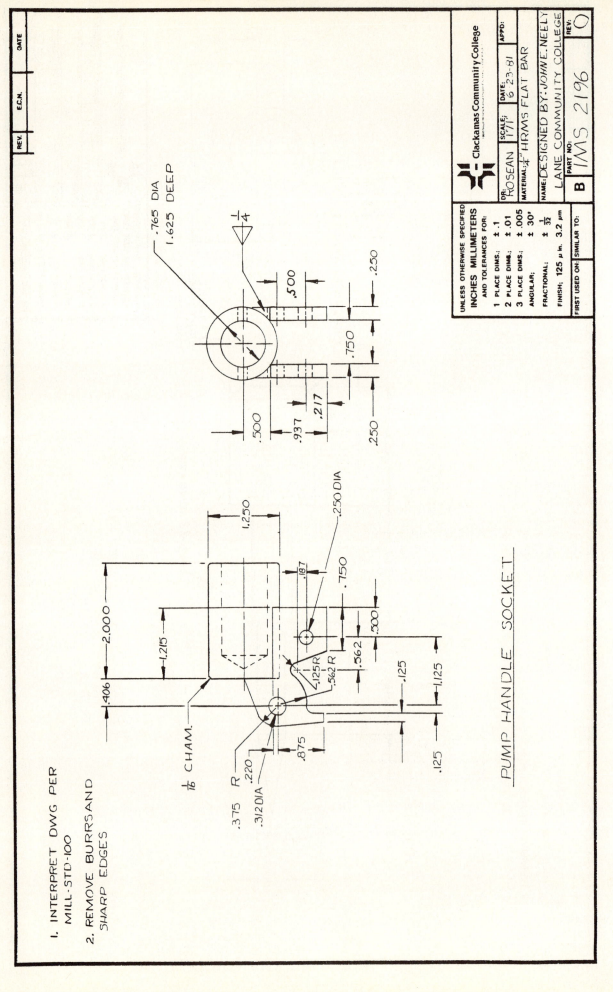

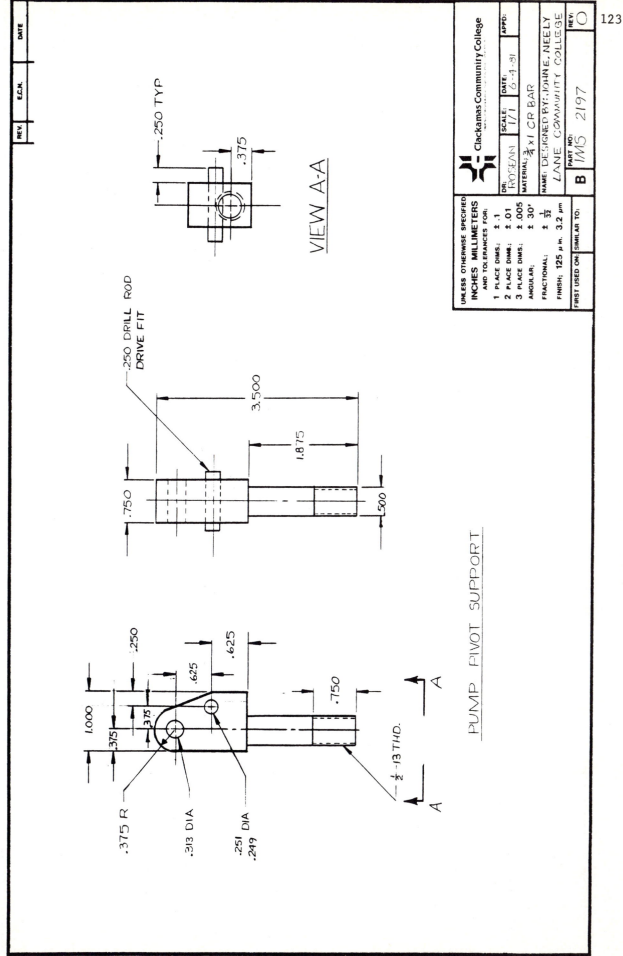

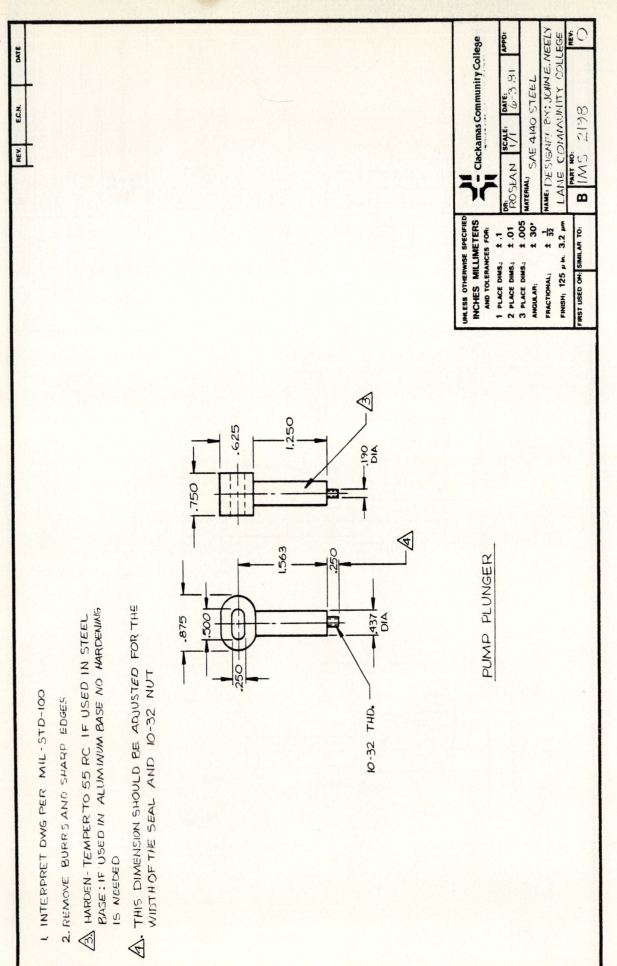

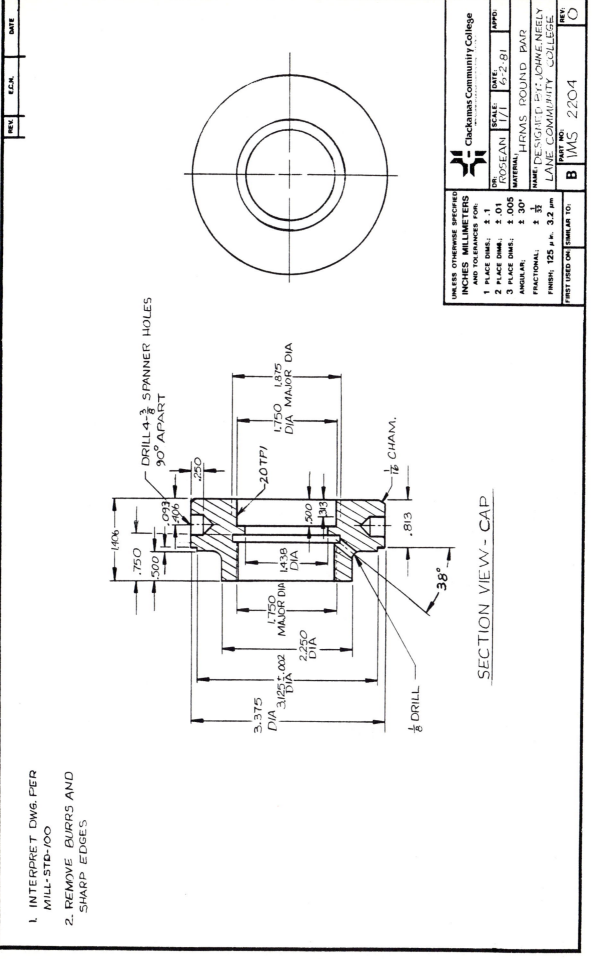

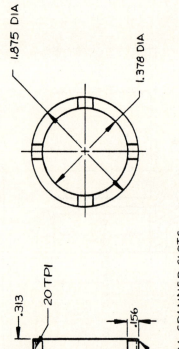

PACKING NUT

1. INTERPRET DWG PER MILL-STD-100
2. REMOVE BURRS AND SHARP EDGES

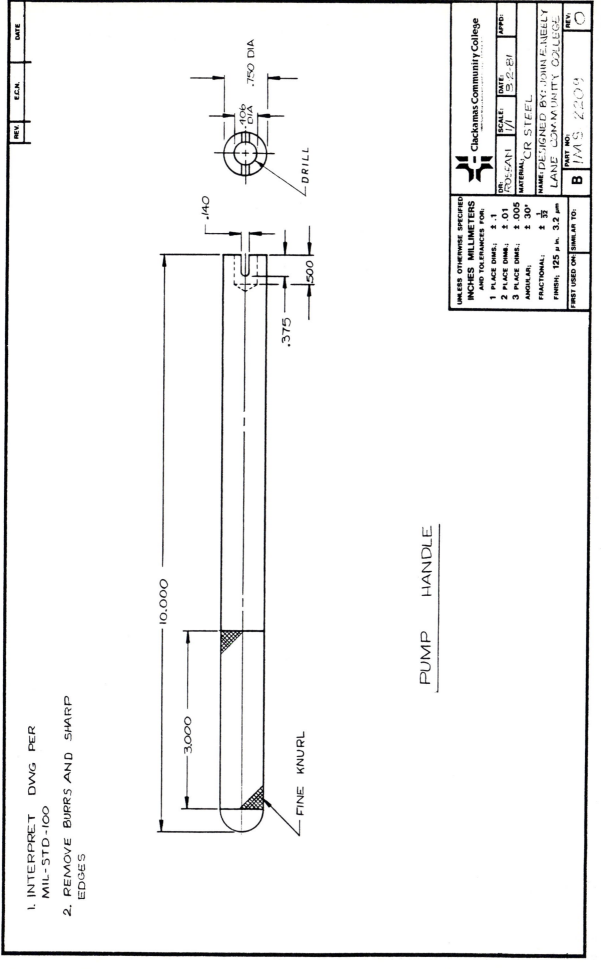

MACHINE TOOL PRACTICES

Name _____ Date _____

PROJECT 15. VEE BLOCKS

<u>Project Evaluation</u> (To be filled out by the instructor):

	Grade	
	Letter	Percent
1. Follows drawing (dimensions and tolerances)	_____	_____
2. Machining finishes	_____	_____
3. Mechanism or tool operates satisfactorily	_____	_____
4. General workmanship	_____	_____
Total grade	_____	_____

<u>Comments</u>:

Signed: _____

(Instructor)

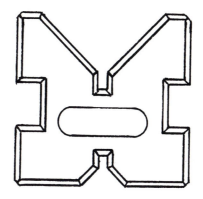

PROJECT 15. VEE BLOCKS

Objectives

1. Use a horizontal mill to square a block and cut vees.
2. Learn heat treating processes for hardening and tempering alloy tool steel.
3. Learn to use Rockwell and Brinell Hardness Testers.
4. Use a precision surface grinder to square and finish vee blocks.
5. Learn to use precision measuring techniques on a surface plate.

Outline for Study

Prior to starting each procedure for this project, study and complete Post-Tests for:

1. Milling the blocks: Refer to Section K, Units 1 through 7.
2. Hardening the blocks: Section D, Units 4 and 5; also refer to Unit 3 for hardening and tempering procedures.
3. Grinding the blocks: Section C, Unit 6; Section N, Units 1 through 6.

Procedures

Begin the procedures for this project by completing the following worksheets.

Worksheet 1. Milling the Vee Blocks
   Exercise 1. Indicating in the Machine Vise
   Exercise 2. Surfacing a Workpiece after the Vise Has Been Indicated In
   Exercise 3. Squaring a Workpiece in a Horizontal Mill
   Exercise 4. Squaring the End of a Workpiece after the Sides Have Been Squared
   Exercise 5. Machining a V-Shaped Part in the Horizontal Mill
Worksheet 2. Hardening and Tempering the Vee Blocks
Worksheet 3. Testing the Vee Blocks
   Exercise 1. Rockwell Hardness Testing
   Exercise 2. Brinell Hardness Testing
Worksheet 4. Surface Grinding the Vee Blocks
   Exercise 1. Care of Gage Blocks and Wringing Practices

WORKSHEET 1. Milling the Vee Blocks

Material

3 inches of 2 x 2 in. SAE 4140 HR bar stock

Procedure

Mill two vee blocks as shown in Figure 1. Also see drawing on page 141.

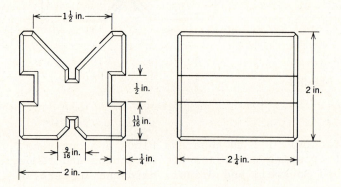

Figure 1.

1. When you have cut off the 5 in. length of alloy steel, leave it full length for squaring operations. The block will be sawed in half later to make two blocks.
2. Deburr the material and prepare the horizontal milling machine as follows.

EXERCISE 1. Milling the Vee Blocks

The steps necessary to indicating-in (aligning) the milling machine vise. Note: This exercise is not needed if the mill vise you are using is aligned to the table with keys.

1. Attach a dial indicator to the arbor or overarm.
2. Position the indicator on the solid jaw of the vise.
3. Move the table by hand so the vise jaw moves past the indicator.
4. Adjust the position of the vise until the indicator readout is zero at each side of the vise jaw.
5. Follow the sequence for squaring a vise in Section K, Unit 5, in the textbook.

EXERCISE 2. Indicating-In the Machine Vise

The steps necessary to surfacing the workpiece. Follow the sequence in Section K, Unit 5, in the textbook.

1. Secure the workpiece in the vise.
2. Install a plain milling cutter (slab mill) that is wider than the workpiece.
3. Position the cutter.
4. Set the speed and table feed.
5. Turn on the cutting fluid and take the first roughing cut.

EXERCISE 3. Squaring a Workpiece in a Horizontal Mill

The steps necessary for squaring a workpiece in a milling machine after one flat surface has been obtained. Follow the sequence for squaring a workpiece in Section K, Unit 5, in the textbook.

1. Deburr the workpiece and clean the vise so it will have an accurate reference surface.
2. Place the deburred flat surface against the solid jaw.
3. Approximately equal amounts should be taken on all sides since hot rolled bars tend to have a decarburized "skin" which is usually about $\frac{1}{32}$ in. deep.

   Decarburization causes the surface to remain soft after the part is hardened.
4. Be careful <u>not</u> to cut to the dimensions given in Figure 1. Those are finished dimensions which will later be obtained by precision grinding. You must leave about 0.030 in. overall for finishing, 0.015 in. on each side. Slight errors in the milling operation (up to 0.003 in. out of square or taper) are permissible since the grinding operation will correct these errors. If the error is greater than this amount, it may still be possible to "save" the piece. Ask your instructor.
5. Remove the part, deburr, and clean the vise and parallels.
6. Check the part with a square and check all of the dimensions with a micrometer.

EXERCISE 4. Squaring the End of a Workpiece after the Sides Have Been Squared

The steps necessary to squaring the ends of a workpiece after the sides have been squared. Follow the sequence for milling the ends of a workpiece in Section K, Unit 5, in the textbook.

1. Place the 5 inch long block in the cutoff saw and cut it into two equal halves. This leaves enough material on each block for milling the ends square.
2. Place the workpiece in the milling vise with the end up and align it with a square.
3. Take a clean-up cut with the slab mill.
4. Reverse the workpiece and mill the opposite end to 0.030 in. over finish size.
5. Remove the workpiece and deburr.

EXERCISE 5. Machining a V-Shaped Part in the Horizontal Mill

The steps necessary to machine a vee-shaped part in the horizontal milling machine.

1. Position the mill vise so one block can be gripped on the ends while supported in a vee-block that is shorter than the workpiece (Figure 2). Make sure the workpiece is well seated against the vee-block. A 90-degree vee can be milled in this manner.
2. Mount a large side-cutting milling cutter on the arbor. The cutter must be large enough to cut to the layout line without allowing the arbor to touch the workpiece.
3. Cut to the layout line of the vee (Figure 3). Several passes should be used since this is not a good setup for heavy cuts. In conventional milling, the cutter could possibly lift the workpiece out of the vise if a full depth cut were made.
4. Repeat the procedure on the second block.
5. Mill the smaller vees on the opposite sides of both blocks.
6. Change to a ½ inch wide side-cutting cutter and mill the ½ inch slots on the sides. These may be cut in one pass in a No. 2 or larger horizontal mill. The blocks may be held in the vise on parallels in much the same way as when the vees were cut. Place the blocks as far down as possible into the vise jaws.

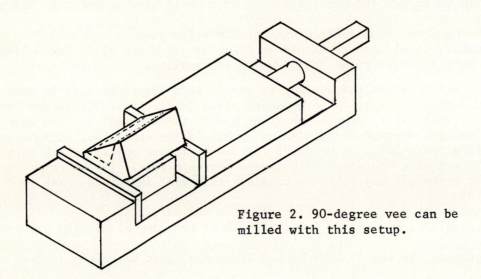

Figure 2. 90-degree vee can be milled with this setup.

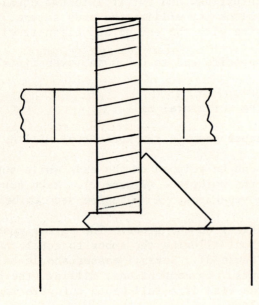

Figure 3. A side-cutting milling cutter is required for this operation.

7. Slitting cutters $\frac{3}{16}$ and $\frac{1}{8}$ in. wide should be used to cut the slots in the bottom of the vees.
8. All edges should be evenly chamfered with a file as shown on the vee block drawing on page 141.
9. A slot should be milled $\frac{3}{8}$ in. wide and $\frac{1}{16}$ in. deep in both blocks so that a name or initials can be stamped on it for identification purposes. This operation should be done on a vertical milling machine.
10. Machines and floor areas should no be cleaned and tooling put away.

WORKSHEET 2. Hardening and Tempering the Vee Blocks

Materials

    Two heat treating furnaces
    Oil quenching bath
    Tongs and safety equipment
    Previously machined set of SAE 4140 steel vee blocks

Procedure

(a) Harden the vee blocks.
(b) Temper the vee blocks to Rc 48 to 52 (HB 470 to 514).

Note: Soaking times for hardening and tempering operations are based on the smallest cross section of the part. About $\frac{3}{4}$ to 1 in. may be considered as the least cross section of this vee block.

1. Set the furnace thermocouple control for the correct hardening temperature for the SAE 4140 steel, which is 1550°F (843°C).
2. Set a second furnace to the correct tempering temperature and turn it on. Consult the SAE 4140 mechanical properties chart (Figure 4) to determine the temperature to obtain a draw temper hardness of Rc 48 to 52 (HB 470 to 514). Note on Figure 4 the 4140 is HB 601 (approximately Rc 60) in the as-quenched condition. In order to reduce this hardness to about HB 500, the temperature would need to be set at about 550°F (288°C) on the tempering furnace. See Table 1 on page 217 in the textbook.
3. The vee blocks should be protected from decarburization by any of several means. A simple method is to first warm the blocks to several hundred degrees (about 400°F), sufficient to melt the boric acid powder. Lightly sprinkle boric acid powder on the hot vee blocks on all sides and place them in the furnace on a fire brick; use tongs.

4. After the vee blocks are the same color as the furnace, allow a soaking time of 1 hour per inch of least cross section (about $\frac{3}{4}$ to 1 in., or $\frac{3}{4}$ to 1 hour).
5. Using face shield and gloves, heat the tongs on the gripping end. Remove one vee block from the furnace, close the door, and quickly plunge the block into the oil bath, agitating the block until it has cooled. It should still be warm to the touch. Repeat the process with the second vee block.
6. Place the two warm vee blocks immediately into the tempering furnace and hold them at that temperature for $\frac{1}{2}$ hour.
7. Remove the blocks and allow them to cool in air.
8. Check for hardness. Because of possible decarburization, the readings may be low. Check again after surface grinding. See Worksheet 3 for hardness testing.
9. Is your vee block the same hardness that the mechanical properties chart indicated it would be at your selected tempering temperature? If not, what reason can you give for the difference?

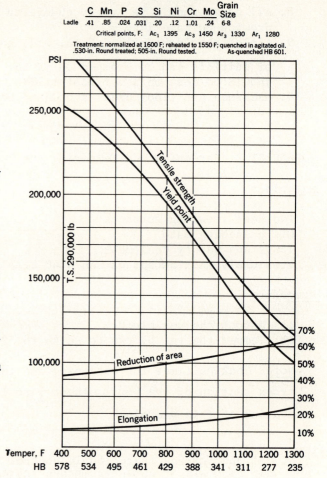

Figure 4. Mechanical Properties chart for SAE 4140 steel (Courtesy of Bethlehem Steel Corporation).

WORKSHEET 3. Testing the Vee Blocks

EXERCISE 1. Rockwell Hardness Testing

<u>Materials</u>

Rockwell hardness tester
Hardened steel
18 gage sheet steel
Nonferrous metal
Tungsten carbide insert
Small part with nitrided steel case

<u>Procedure</u>

(a) Three hardness tests on the standard Rockwell tester.
(b) Two tests on the superficial tester.

1. Before starting, see that the crank handle is forward.
2. Select proper penetrator and insert it in the plunger rod.
3. Place the proper anvil on the elevating screw.
4. Select the proper weights.
5. Place the specimen on the anvil.

6. Raise the specimen into contact with the penetrator by turning the capstan handwheel clockwise. Continue motion until the small pointer is near the dot. Continue until the larger pointer is in a vertical position. The minor load is now applied.
7. Turn the bezel of the dial gage until the "SET" line is directly behind the large pointer.
8. Release the weights (major load) by tripping the crank handle clockwise. Do not force the crank. Allow it to come to rest.
9. When the large pointer has come to rest, return the crank handle to the starting position. This removes the major load; the minor load is still applied.
10. Read the scale letter and Rockwell hardness number from the dial gage.
11. Remove the minor load by turning the capstan handwheel counterclockwise to lower the elevating screw and specimen.
12. Remove the specimen. Repeat the test procedure in one or more locations on the test piece.

Material	Surface Condition	Rockwell Scale	Penetrator	Weight	Hardness Reading 1 2 3 Average
1. Nonferrous metal					
2. Hardened steel					
3. Tungsten carbide insert					
4. 18 gage sheet steel					
5. Nitrided steel part					

Record your results and show then to your instructor.

EXERCISE 2. Brinell Hardness Testing

Materials

Brinell hardness tester
Heat treated steel

Nonferrous material (aluminum or copper)
Soft steel

Procedure

(a) Make three hardness tests.

1. Place the specimen on the anvil.
2. Raise the specimen until it touches the ball.
3. Apply the load for 30 seconds. Release.
4. Remove the specimen from the tester and measure the diameter of the impression.
5. Determine the Brinell Hardness Number.
6. Check for hardness in two or three locations on the specimen.

Material	Surface	Weight	Reading
1. Nonferrous metal			
2. Hardened steel			
3. Soft steel			

Record your results and show them to your instructor.

Testing the Blocks

1. After completing the Rockwell and Brinell hardness testing exercises (on both or whichever tester is available to you), you are now ready to test your vee block.
2. Take three tests in different locations and record the results. If the tests do not average 48 to 52 RC, try again after some of the surface has been removed by surface grinding. If the blocks did not get hard for some reason, they may be rehardened; but first they must be normalized as explained in Section D, Unit 4.

WORKSHEET 4. Surface Grinding the Vee Blocks

Materials

Surface grinder and accessories
A set of vee blocks previously machined and heat treated

Procedure

1. Grind all sides square to each other within plus or minus 0.0005 in.
2. Hold the decimal dimensions to within 0.0005 in. and the fractional dimensions within plus or minus $\frac{1}{64}$ in. Follow the steps given in Section N, Unit 6, pages 673 to 679 in your textbook.
3. Obtain a fine finish (about 16 to 20 microinch) on all surfaces with the final grinds. Check with your instructor on measurement, squareness, and finish from time to time as you progress with this project.

EXERCISE 1. Care of Gage Blocks and Wringing Practices

Materials

Shop set of gage blocks	Height gage
Clean solvent	Surface plate
Lint free tissue	Insulated forceps (if available)
Camel's hair brush	Shop cloth or lint free cloth
Gage block tray	Gage block preservative
Dial test indicator	Conditioning stone (granite or ceramic)
	(Be sure that this is the correct item!)

Procedure

1. Select two gage blocks from each size series and lay them on the tray which has been covered with a clean shop towel or lint free cloth. Select two wear blocks.
2. Moisten a lint free tissue in clean solvent and clean the surface of the conditioning stone. Also clean the surfaces of the gage blocks to be wrung.
3. Dry the surfaces of the conditioning stone and the blocks to be wrung.
4. After cleaning, place the blocks on their sides to reduce the amount of dust settling on the cleaned surfaces.
5. If more than a few seconds elapse after cleaning and before wringing, dust the surface to be wrung with the camel's hair brush.
6. Bring the surfaces into contact and wring the blocks using the procedure described in the textbook, page 173.
7. If the conditioning stone must be used, first dust the stone with the camel's hair brush. Deburr the block with a single back and forth pass over the stone.
8. Wring the remaining blocks together, using a wear block at each end of the stack.
9. Make a stack 2¼ in. high to check the length of the vee blocks.
10. Set up a 0.0001 in. discriminating dial test indicator on a height gage and place them on a surface plate. Compare the height of the vee blocks to the gage block stack.
11. Repeat the process in steps 9 and 10 to check the 2 inch dimensions of the vee blocks.
12. Disassemble the blocks by sliding them apart. Spread them apart on the clean shop towel or tissue.
13. If you do not use gage blocks for measuring the vee blocks, disregard steps 9 through 11.
14. Turn in the grading sheet and the completed vee blocks to your instructor for evaluation.

VEE BLOCK

VEE BLOCK DRAWING

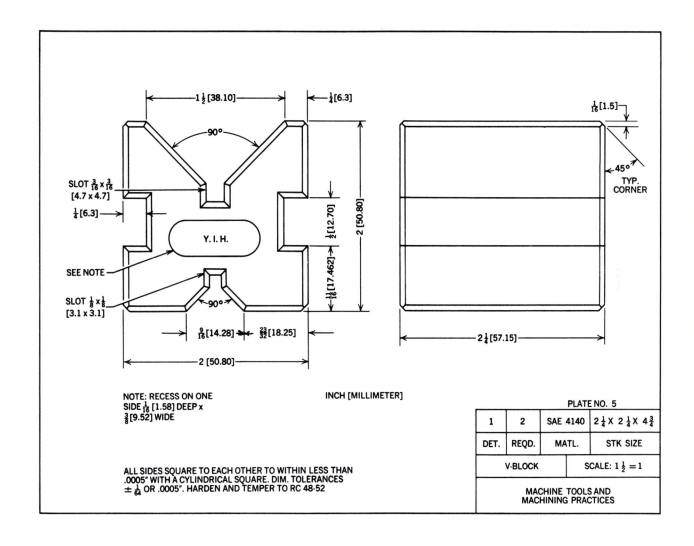

MACHINE TOOL PRACTICES

Name _____ Date _____

PROJECT 16. PRECISION VISE

<u>Project Evaluation</u> (To be filled out by the instructor)

	Grade	
	Letter	Percent
1. Follows drawing (dimensions and tolerances	_____	_____
2. Machining finishes	_____	_____
3. Mechanism or tool operates satisfactorily	_____	_____
4. General workmanship	_____	_____
Total grade	_____	_____

<u>Comments</u>:

Signed: _____
(Instructor)

MACHINE TOOL PRACTICES 145

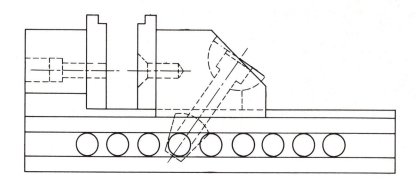

PROJECT 16. PRECISION VISE

Objectives

1. Learn to square a block and to machine steps and grooves in the block on a horizontal mill.
2. Be able to drill and ream holes and machine slots on a vertical mill.
3. Be able to precision grind for perpendicularity and dimension.

Outline for Study

Prior to starting each procedure for this project, study and complete Post-Tests for:

1. Machining the block: Refer to Section K, Units 5, 6, and 7; Section C, Unit 8.
2. Milling the slot and cavity in the vise body: Section J, Unit 5.
3. Making the movable jaw and parts: Refer to Section J, Units 3 and 4.
4. Hardening and tempering the vise parts: Refer to Section D, Unit 3.
5. Grinding the vise parts: Section N, Unit 7; also refer to Units 1 through 6.

Procedures

Begin the procedures for this project by completing the following worksheets.

Worksheet 1. Machining the Block for the Vise Body
Worksheet 2. Machining Holes in the Vise Body
Worksheet 3. Milling the Slot and Cavity
Worksheet 4. Making the Movable Jaw and Parts
Worksheet 5. Hardening and Tempering the Vise
Worksheet 6. Grinding the Vise Parts

WORKSHEET 1. Machining the Block for the Vise Body

Materials

$6\frac{1}{4}$ inches of $2\frac{1}{2}$ x $3\frac{1}{4}$ in. SAE 4140 steel

Procedure

1. Set up the block in a horizontal milling machine as explained in Section K, Unit 5, pages 579 to 582 in your textbook and mill it square and to the oversize

(to allow for grinding) dimensions shown in Figure 1.

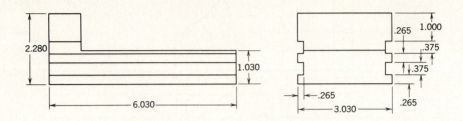

2. Take the steel block that has been squared and machined to length. Make a layout of the grooves and the step to be machined. See Drawings I and II at the end of this project. Follow the procedures for side milling in Section V, Unit 6, of your textbook.
3. Mount a vise on the milling machine table and align the solid jaw so it is square to the column face.
4. To machine the large step on the project, mount the workpiece in the vise. Support the workpiece on parallels so that about $1\frac{5}{16}$ in. extends above the vise top. The step is to be $1\frac{1}{4}$ in. deep; the extra $\frac{1}{16}$ in. is to clear the vise jaws.
5. Select a side milling cutter with a diameter large enough to make a cut $1\frac{1}{4}$ in. deep.
6. Mount the cutter on the arbor so that the cutting pressure will be against the solid jaw. Check for sufficient clearance between arbor support and vise. Check the position of the cutter on the arbor by moving the table cross feed to see if the full width of the cut can be made.
7. Select and set the cutter speed. Use the formula

$$RPM = \frac{CS \times 4}{D} = \underline{\hspace{2in}}$$

8. Check that the spindle rotation is correct.
9. Calculate the feedrate to use and set it on the machine. The feedrate equals feed per tooth x number of teeth x RPM.
10. Set the depth of cut. One or two roughing cuts may be required and a finishing cut will also be needed. The final roughing cut should leave about .030 in. for the finish cut. Note that on Figure 1, the finish dimension is .030 in. oversize.
11. Move the table to set the width of the cut to $\frac{1}{2}$ in. The width of the cut should be less than the width of the cutter.
12. Tighten the knee and cross slide locking clamps.
13. Turn on the coolant.
14. Start the cut manually, then engage the power feed (Figure 2). Observe the cutting operation carefully and be prepared to disengage the power feed at the first sign of trouble.
15. When the cut is completed, return the table to its starting position.
16. Move the table into position to take another $\frac{1}{2}$ in. wide cut.
17. Repeat taking cuts until you are within $\frac{1}{4}$ in. of your layout lines. Then stop the spindle and measure the width of the remaining section. Calculate the amount of oversize. Adjust the machine table by the amount of oversize using the cross feed micrometer collar.
18. Make this cut. Set the cross feed micrometer collar to zero. This is also the reference point for the finishing cut. This completes the roughing cut.

Figure 2. (Lane Community College). Figure 3. (Lane Community College).

19. Return the machine table to the starting point for the first cut. Set the depth of cut for the finishing cut.
20. Make the finishing cuts in the same manner as the roughing cuts, except that you <u>stop the cutter</u> before returning it over the finished surface. Returning a revolving cutter over a surface that was just machined will leave cutter marks.
21. After completing the final cut, stop the machine. Remove the chips and deburr the workpiece. Measure the workpiece while it is still clamped in the vise. If another cut is needed, it should be made next. This completes the milling of the step. Remove the workpiece from the vise.
22. Turn the vise and align the solid jaw parallel with the table.
23. Clamp the workpiece with the long outside (bottom) surface against the solid jaw of the vise as shown in Figure 3.
24. Select a sharp side milling cutter to machine the grooves. Exchange the cutter on the arbor with the new cutter.
25. Position the cutter on the side of the workpiece by using a paper strip. Zero the cross feed micrometer collar.
26. Move the cutter over the groove location using the micrometer collar. The cutter should now be aligned with the layout lines.
27. Set the depth of cut. Only one cut is required for making the full groove depth.
28. Calculate and set the speed for this cut.
29. Calculate and set the feedrate.
30. Turn on the coolant and start the cut manually, then engage the power feed.
31. At the completion of this cut, turn off the feed, coolant, and spindle, and return the table to its starting position.
32. Move the table the required distance to align the cutter for the second groove. Use the micrometer collars to adjust the machine to get accurate locations.
33. Cut the second groove.
34. Stop the machine, then return the table to its starting position.
35. Clean off the chips, deburr the just machined grooves, and remove the workpiece from the vise.
36. Clean the vise and workpiece. Reposition the workpiece in the vise with the grooved side down and with the same side against the solid jaw as used for the last operation. Clamp the workpiece securely.

37. Positioning of the workpiece under the cutter should be done with the aid of the micrometer adjustments. The grooves on the top side should be exact duplicates of the grooves on the bottom side. Remember to remove the backlash that exists when you change the rotation of any of the feedscrews.
38. Cut both grooves.
39. The undercut slot on the jaw for grinding relief should now be cut, using a $\frac{1}{8}$ in. slitting saw. A cutter with a radius would be preferred as a round notch is less likely to crack during heat treatment operations.
40. Clean and deburr the workpiece and remove it from the vise. Clean the machine.
41. This completes the horizontal milling machine work on this workpiece.

WORKSHEET 2. Machining Holes in the Vise Body

Materials

Vertical milling machine and accessories

Procedure

1. Lay out the vise body you have just milled to the oversize dimensions shown in Figure 4.

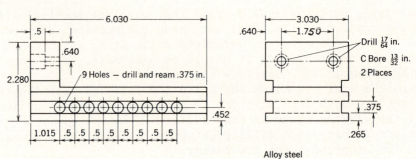

Figure 4. The holes in the vise body.

2. Clamp the workpiece in the vise on a vertical mill table with a side up. Set up a drill chuck in the spindle.
3. Using an edge finder, locate the spindle to the large end of the vise body and center it to the groove in which the holes are to be drilled.
4. Determine the distance to the first hole from Drawing II and add .015 in. for oversize. Do not forget to subtract one-half the diameter of the edge finder.
5. Spot drill, drill under the reamer size, and ream one hole at each location. Then move to the next hole. Drill and ream only halfway through the block. See Figures 5 and 6.
6. When one side has been finished, turn the block over and set up again, and repeat the previous steps.
7. Now turn the vise body on end as shown in Figure 7 and drill and counterbore as shown in Drawing II.

Figure 5. Drilling $\frac{11}{32}$ in. diameter holes in the vise body (Lane Community College).

Figure 6. Reaming $\frac{3}{8}$ in. diameter holes in the vise body (Lane Community College).

Figure 7. Drill and counterbore holes in the solid jaw of the vise body (Lane Community College).

WORKSHEET 3. Milling the Slot and Cavity

Materials

Vertical milling machine and accessories

Procedure

1. Fasten the workpiece in the vise bottom up; use parallels so the workpiece extends above the vise for measuring while it is being machined.
2. Use a $\frac{7}{16}$ in. diameter center cutting end mill to rough out the $\frac{1}{2}$ in. wide slot (Figure 8).

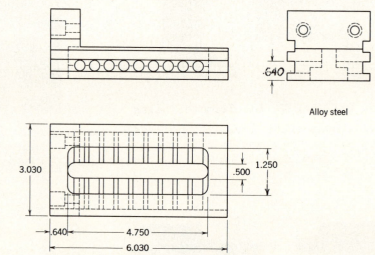

Figure 8.

3. Calculate the RPM.
4. Calculate the feedrate.
5. Align the end mill over the slot and lock all machine axes that will not have any movement during the cutting operation.
6. Rough out the slot; use coolant while cutting.
7. Change to a $\frac{1}{2}$ in. end mill.
8. Make a $\frac{1}{8}$ in. deep trial cut, then measure the location of the slot in the workpiece. Make any necessary adjustments to put the slot into the center of the workpiece.
9. Finish machine the $\frac{1}{2}$ in. slot.
10. Rough out the cavity (Figure 9), leaving $\frac{1}{32}$ in. for the finish cut all around. Keep the cutting pressure against the tool by using conventional (up milling) techniques.
11. Measure the depth of the cavity with a depth micrometer. Then make the depth adjustment necessary for the finish cut.
12. Make a cut with the cutter just touching the side of the cavity, measure the distance with a micrometer, then adjust for one final cut the correct distance in from the outside edge.
13. Machine each side of the cavity, repeating the operations performed in step 12.
14. Deburr all edges.
15. This completes the machining of the vise body prior to heat treating and finish grinding.

Figure 9. Milling out the cavity (Lane Community College).

Figure 10. Using a shell mill to machine the angle (Lane Community College).

WORKSHEET 4. Making the Movable Jaw and Parts

Materials

$3\frac{1}{4}$ inches of $1\frac{1}{2}$ x 2 in. SAE 4140 HR steel

$6\frac{1}{2}$ inches of $\frac{5}{16}$ x $1\frac{3}{4}$ in. spring steel or SAE 4140 steel

2 inches of 1 in. dia. SAE 4140 steel round

Procedure for the Movable Jaw

1. Square the block in either a horizontal mill or by using a carbide face mill in a vertical milling machine. Leave it 0.030 in. oversize.
2. Clean and deburr the block and lay out for drilling and milling as specified in Drawing III.
3. Drill and tap the $\frac{1}{4}$-28 screw holes.
4. Set the milling head to 45 degrees and mill the angular surface as shown in Figure 10.
5. With the same setup, drill the $\frac{3}{8}$ in. hole.
6. Set up a 1 in. dia. ball end cutter and machine the ball socket.
7. Turn the part in the vise so the bottom side is up and mill the recess. Remember to make it 0.030 in. less in width than the finish dimension.
8. Make the undercuts as shown on Drawing III.
9. Using a $\frac{3}{8}$ in. end mill with the same setup, elongate the $\frac{3}{8}$ in. drilled hole to make a slot that will be vertical to meet the ball socket on one end, and angular on the other. This slot will allow the $\frac{3}{8}$ in. clamping screw free movement so it can clamp at different angles.
10. This completes the milling on the movable jaw. It can now be hardened and tempered along with the vise body.

Procedure for the Jaw Insert

1. Saw off $6\frac{1}{2}$ inches of the tool steel for the jaw inserts and mill them to the dimension shown in Drawing IV on the edges, allowing 0.015 in. for grinding. If chrome spring steel is used, remember it is prone to work hardening so the cutting speed must be very low.
2. Cut the piece in two, and mill the ends, leaving 0.030 in. oversize.
3. Lay out for the holes. Drill and countersink on one and drill and tap on the other.
4. The jaw inserts may be heat treated along with the vise body and movable jaw with the same settings for hardening and tempering.

Procedure for the Swivel Nut

1. The nut may be turned on a lathe as shown in Drawing IV.
2. Place the turned piece in the vertical mill and cut the angles.
3. The nut may be hardened along with the other parts, but it should be tempered at about 650°F (343°C) so that it will be tougher causing it to resist bending or breaking.

Procedure for the Ball Washer

1. The ball washer can be turned on a lathe by first drilling and counterboring the square end. Make a facing cut. Chamfer.
2. Use a ball turning attachment or free hand the radius toward the chuck, leaving a stem at least $\frac{3}{8}$ in. diameter. Check frequently with a radius gage. Finish with a file and abrasive cloth.
3. The ball washer need not be hardened, but left in the as-rolled condition. A $\frac{5}{16}$-24 by 2 in socket head capscrew will also be needed to complete the jaw parts.

WORKSHEET 5. Hardening and Tempering the Vise

Materials

Heat treating furnaces and accessories; protective clothing, gloves, face protection

Procedure

1. Hardening and tempering will be done in much the same way as that for the vee blocks and according to methods explained in Section D, Unit 3. Two furnaces should be used, one for hardening and one for tempering, so no delay will take place between the two operations.
2. Some means such as using a light coating of boric acid powder on the parts should be used to prevent decarburization and scaling.
3. The large parts should be put in a cold furnace and brought up to the hardening temperature. The small parts can be put in a hot furnace.
4. Holes in the larger parts should be plugged with steel wool, and a wire fastened in a convenient location so the part can be easily picked up in the correct orientation for quenching in the oil.
5. The quenching oil should be slightly warm to the touch. Note: A quench crack can instantly ruin a part that has many hours of work in it. Quench cracks are caused most often by a hardening temperature that is set too high. They can be caused at the intersection of a large mass and a small mass if they are not quenched correctly. In this case, arrange for the large end of the vise body to enter the quench bath first; do not hesitate -- plunge it quickly under the oil and agitate it. If the quench bath is quite cold, the shock may cause a crack. For most small parts, a cold quench is acceptable, but with more massive pieces, the quench bath should be warmed. These parts seldom develop a quench crack and, when they do, it can always be traced to an extreme violation of good practice.
6. As soon as the parts have been quenched and cooled so they are still warm to the touch, they should be wiped off (to remove oil) and put directly in the tempering furnace and kept there for about one-half hour.
7. When the parts have been cooled, they should be tested for hardness.

WORKSHEET 6. Grinding the Vise Parts

Materials

Surface grinder and accessories; surface plate and precision measuring devices

Procedure

1. Methods of surface grinding as explained in Section N, Units 1 through 7 will be used to grind the vise parts. The procedures will be much the same as those used

for the vee blocks.
2. The bottom side of the vise body should be first ground by supporting the body on magnetic parallels.
3. Clean the part and wipe off the magnetic table. Place the vise right side up on the table, crosswise to the travel so the jaw can be ground in the same setting. The vise must be set square to the travel.
4. Grind the horizontal surface with a straight wheel. Change to a dish wheel and clean up the jaw face.
5. Replace the straight wheel and clean up grind all sides, leaving them oversize.
6. Use the squaring technique with a precision angle plate, surface gage, and dial indicator that you used on the vee blocks.
7. When the vise body and movable jaw are relatively square, check them for squareness on a surface plate with a precision cylindrical square. They should be within plus or minus .0005 in perpendicularity. Of course, they must still be oversize. Have your instructor observe this check.
8. When they are square within the acceptable tolerance, finish grind the grooves on the side of the body with a $\frac{1}{4}$ inch wide straight wheel.
9. True up a straight wheel for finishing. Finish grind the vise body and the movable jaw to approximately 16 microinches and to a dimensional tolerance of plus or minus .0005 in. Note: The movable jaw must have a sliding fit on the vise body; not over 0.003 in. clearance, but free to move.
10. The jaw inserts may now be surface ground to size, and the notches and chamfers ground.
11. ALL sharp corners and edges must now be slightly chamfered with a fine stone by hand. These sharp edges can cause a severe cut unless they are dulled.
12. Clean and oil all parts. You may now assemble the precision vise.
13. Turn in the grading sheet and the completed precision vise to your instructor for evaluation.

PRECISION VISE: DRAWING I

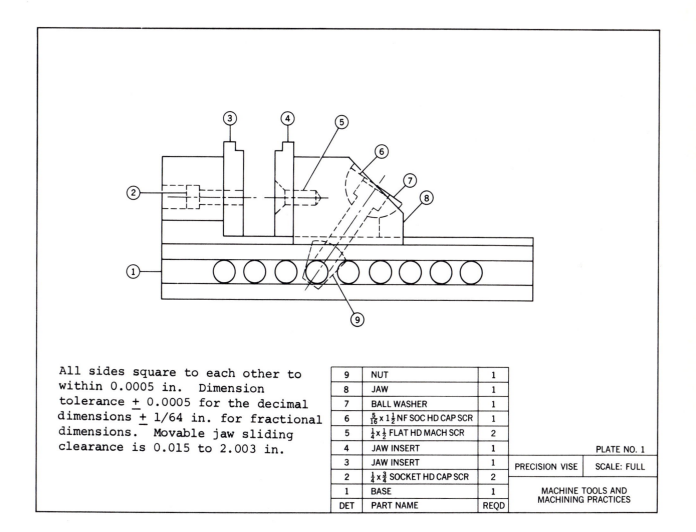

All sides square to each other to within 0.0005 in. Dimension tolerance ± 0.0005 for the decimal dimensions ± 1/64 in. for fractional dimensions. Movable jaw sliding clearance is 0.015 to 2.003 in.

DET	PART NAME	REQD
9	NUT	1
8	JAW	1
7	BALL WASHER	1
6	$\frac{5}{16} \times 1\frac{1}{2}$ NF SOC HD CAP SCR	1
5	$\frac{1}{4} \times \frac{1}{2}$ FLAT HD MACH SCR	2
4	JAW INSERT	1
3	JAW INSERT	1
2	$\frac{1}{4} \times \frac{3}{4}$ SOCKET HD CAP SCR	2
1	BASE	1

PLATE NO. 1
PRECISION VISE — SCALE: FULL
MACHINE TOOLS AND MACHINING PRACTICES

PRECISION VISE: DRAWING II

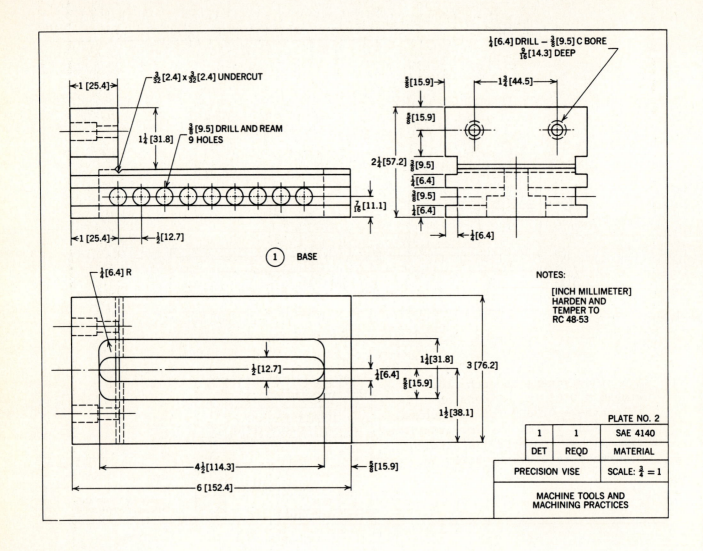

PRECISION VISE: DRAWING III

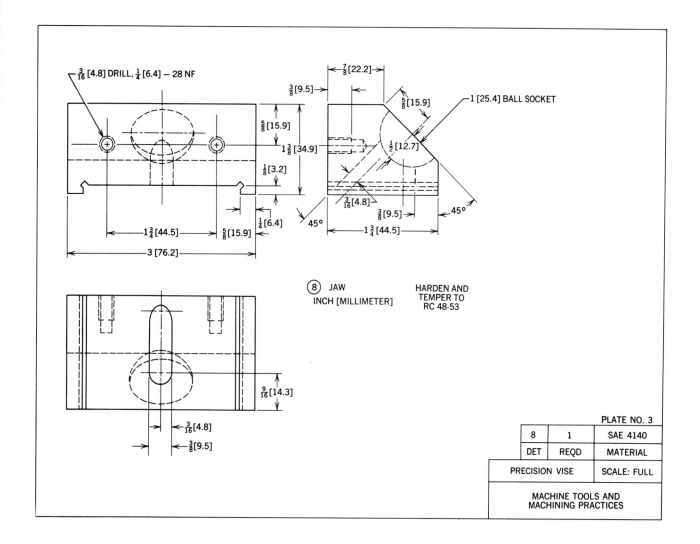

PRECISION VISE: DRAWING IV

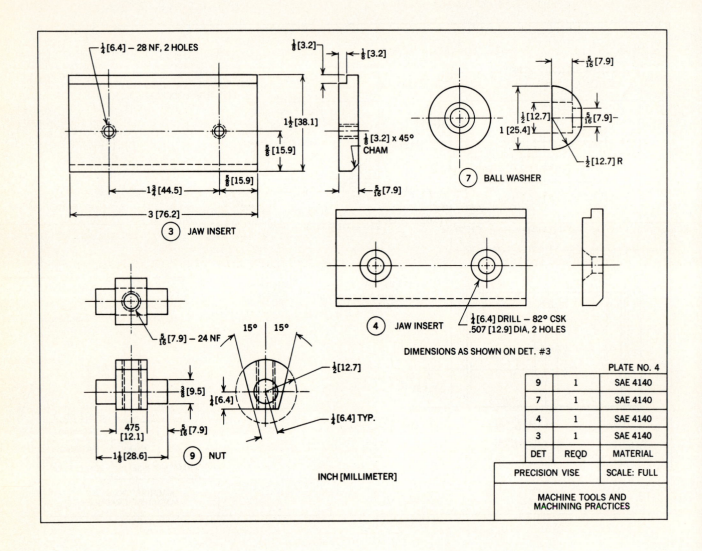

MACHINE TOOL PRACTICES

Name _____ Date _____

PROJECT 17. SPUR GEAR

Project Evaluation (To be filled out by the instructor):

	Grade	
	Letter	Percent
1. Follows drawing (dimensions and tolerances)	_____	_____
2. Machining finishes	_____	_____
3. Mechanism or tool operates satisfactorily	_____	_____
4. General workmanship	_____	_____
Total grade	_____	_____

Comments:

Signed: _____
(Instructor)

MACHINE TOOL PRACTICES 161

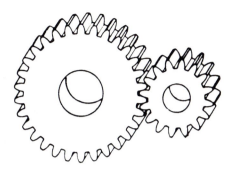

PROJECT 17. SPUR GEAR

Objectives

1. Learn how to prepare a gear blank for cutting a spur gear.
2. Learn to set up for gear cutting.
3. Use an indexing head.

Outline for Study

Prior to starting each procedure for this project, study and complete Post-Tests for:

1. Preparing the gear blank: Section M, Units 1 and 2.
2. Setting up for gear cutting: Section M, Units 3 and4. Refer Section K, Units 4, 5.
3. Using an indexing head: Section L, Unit 1.

Procedures

Begin the procedures for this project by completing the following worksheets.

Worksheet 1. Preparing the Gear Blank
   Exercise 1. Reaming in the Lathe
   Exercise 2. Press a Mandrel into a Bore
Worksheet 2. Set up for Gear Cutting
Worksheet 3. Cutting the Gear

WORKSHEET 1. Preparing the Gear Blank

Material

The size and pitch of the gear should be assigned by your instructor since it may be limited to the available gear cutters. Perhaps the shop needs a particular gear for a machine tool or perhaps for your final term project. The spur gear (shown in Figure 1 in Worksheet 2) can be made with your instructor's approval. The material needed will be subject to these requirements.

Procedure

1. The gear blank diameter is determined by the diametral pitch and the number of teeth on the gear as explained in Section M, Unit 2, of your textbook. Leave the diameter and width slightly oversize so it can be finished while on a mandrel.

2. The bore should be tentatively made a nominal size to fit a mandrel sufficiently large to hold the blank securely, yet at or under the required final bore size. For example, a 2 to 3 inch diameter gear should have at least $\frac{3}{4}$ in. bore and a 5 to 6 inch gear, a 1 in. bore to hold them securely.
3. One way to quickly make this bore in the blank is to ream it to size in the lathe.

EXERCISE 1. Reaming in the Lathe

Material

Lathe   Center drill $\frac{5}{16}$ in. machine reamer

$\frac{11}{16}$ in. drill $\frac{19}{64}$ in. drill $\frac{3}{8}$-16 tap and tap wrench

$\frac{3}{4}$ in. dia. 1 in. long mild steel HR round

$1\frac{1}{4}$ in. dia. $1\frac{1}{2}$ in. long mild steel HR round

Procedure

Drill and ream a hole in a gear blank.

1. Mount the gear blank in a chuck. Face and center drill.
2. Drill through with a correct size drill for reaming. Use cutting oil.
3. Chuck the reamer in a drill chuck or socket in the tailstock and ream through. Use cutting oil.
4. Chamfer the hole on both sides.
5. Check the diameter with a telescoping gage and micrometer. If it is more than 0.001 in. oversize, a standard tapered mandrel may not fit. A nonstandard or adjustable mandrel will work in that case.

EXERCISE 2. Press a Mandrel into a Bore

Materials

Arbor press   Gear or similar part with a bore size to fit the mandrel
High pressure lubricant

Procedure

1. Apply high pressure lubricant to the mandrel and bore.
2. Insert the <u>small</u> end of the mandrel into the bore; it should not go in more than about one-half the length of the mandrel.
3. Place the gear and mandrel on the bolster or support plate of an <u>arbor press</u>, not a powered hydraulic press.
4. Apply pressure on the mandrel until it is firmly seated. If the mandrel is not sufficiently large to tighten in the bore, you must change to a larger mandrel or an adjustable mandrel. A loose mandrel may allow slippage while the gear is being cut, thus ruining the gear.

WORKSHEET 2. Setup for Gear Cutting

Material

　　Mild steel spur gear blank

Procedure

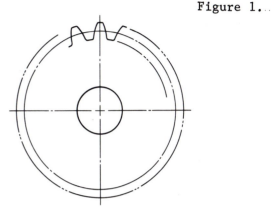

Figure 1.

　　Make the spur gear according to the following data unless you already have a selected or assigned gear to cut.

　　32 tooth gear, $\frac{3}{4}$ in. face
　　12 diametral pitch
　　$14\frac{1}{2}$ in. pressure angle teeth

1. Calculate the gear dimensions required for the previously selected spur gear as noted in Worksheet 1, or alternately, make the spur gear shown in Figure 1.
　　　Outside diameter _____　　　Pitch diameter _____
　　　Whole depth _____　　　　Chordal addendum _____
　　　　　　　　　　　　　　　　　　Chordal thickness _____
2. Assuming that the gear blank is now securely mounted on a mandrel, set up the mandrel between centers in a lathe and finish the blank to size, both on diameter and for width.
3. Set up a dividing (indexing) head on a horizontal milling machine. The head must be equipped with a center and dog plate or arm. Set up the footstock.
4. Place the gear blank and mandrel between centers with a driving dog. The bent tail of the dog must be clamped so it cannot move. The large diameter of the mandrel should be toward the dividing head.
5. The gear blank and mandrel must be parallel to the table surface. This can be checked with a dial indicator and surface gage. Remember, most mandrels are about 0.004 in larger in diameter on one end, so the indicator should read 0.002 in. more on that end (when it is parallel) in that case. Check for the exact difference in diameter on the mandrel you are using and compensate accordingly.
6. Most gear cutters have the total depth of cut marked on the side. Check this number to see if it is the same as your calculated number.
7. Mount the gear cutter on an arbor and set it up on the machine. Set the RPM and feedrate.
8. Center the cutter over the gear blank axis.
9. Zero the knee crank dial at the point at which the cutter touches the gear blank surface.
10. Set the depth for the roughing cut. About 0.030 in. should be left for finishing. Two roughing cuts may be necessary, depending on the rigidity of the setup.
11. Calculate the index movement necessary and set the sector arms to the required number of holes.
12. Mark all of the gear spaces by indexing around the gear and slightly nicking the gear blank with the revolving cutter.
13. Make the first roughing cut, then adjust and set the table feed trip dogs at the beginning and end of the cut.
14. Make all roughing cuts.
15. Set the depth for the finish cut and cut two spaces.

16. Measure the tooth thickness.
17. Complete the finish cut on all spaces.
18. Remove and deburr the gear.
19. Clean the milling machine and accessories.
20. Turn in the grading sheet and the completed spur gear to your instructor for evaluation.

MACHINE TOOL PRACTICES

Name _____ Date _____

PROJECT 18. CYLINDRICAL GRINDING

<u>Project Evaluation</u> (To be filled out by the instructor):

	Grade	
	Letter	Percent
1. Follows drawing (dimensions and tolerances)	_____	_____
2. Machining finishes	_____	_____
3. Mechanism or tool operates satisfactorily	_____	_____
4. General workmanship	_____	_____
Total grade	_____	_____

<u>Comments</u>:

Signed: _____

(Instructor)

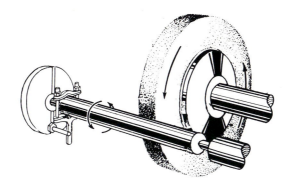

PROJECT 18. CYLINDRICAL GRINDING

Objectives

1. Learn to use the cylindrical grinding machine to make a straight precision cylinder.
2. Be able to produce a precision taper.
3. Be able to grind an internal surface.

Outline for Study

Prior to starting each procedure for this project, study and complete Post-Tests for:

1. Grinding a cylindrical shape: Section N, Units 8 and 9.
2. Grinding external and internal tapers: Refer to Section I, Unit 13.

Procedures

Begin the procedures for this project by completing the following worksheets.

Worksheet 1. Making a Lathe Mandrel; Grinding a Cylinder
Worksheet 2. Grinding a Taper
Worksheet 3. Grinding a Bore

WORKSHEET 1. Making a Lathe Mandrel; Grinding a Cylinder

Material

$7\frac{1}{8}$ in. of 1 in. SAE 4140 steel for a $\frac{3}{4}$ in. mandrel or sufficient material for a selected mandrel size. Standard mandrel dimensions may be found in Machinery's Handbook.

Procedure

1. Turn a mandrel between centers on a lathe as shown in Figure 1. Check with your instructor to see if a different size or another part is to be made.
2. Make the turned diameter 0.015 to 0.030 in. oversize (depending on size).
3. Mill the flats on the ends.
4. Harden and temper to about RC 50.

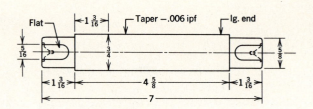

Figure 1. Sketch of tapered lathe mandrel

5. Follow the instructions in your textbook in Section N, Unit 9, for grinding a mandrel. Before beginning the mandrel grinding procedure for grinding the taper given on page 686 in the textbook, make a straight grind.
6. When you have made the setup, before turning on the wheel, have your instructor inspect it.
7. With the wheel and work rotating and the table traverse set, feed the wheel to the work. Infeed the wheel until it just begins to produce sparks.
8. Turn on the coolant.
9. Make several passes, infeeding about 0.001 to 0.002 in. per pass until the surface scale is all cleaned up. Allow the table to traverse without infeed until it sparks out. Retract the wheel and stop the rotation and table traverse.
10. Check the two ends of the mandrel with a "tenth" micrometer. It should be the same within one "tenth."
11. If there is taper, move the table to compensate and regrind until the mandrel is all the same diameter within plus or minus 0.0001 in. The mandrel should still be over the final finish diameter.

WORKSHEET 2. Grinding a Taper

Materials

The mandrel

Procedure

1. Follow the steps given in your textbook on page 689 for grinding a taper. When adjusting the swivel table for the required taper, a dial indicator may be used in a position near the end of the table at a given distance from the pivot point of the table. This distance and the dial indicator reading is a proportional ratio to the ratio of amount to be removed from the mandrel <u>on one side</u> in a given length. For example, in 6 inches the mandrel should taper 0.003 in. in diameter. The table must be rotated so that 0.0015 in. more is removed from one end <u>on one side</u> than is removed on the other. If the table pivot point is 30 inches from the dial indicator, the ratio would be 30 in. to 6 in. or 5:1. Therefore, the dial indicator movement should be 5 x 0.0015 = 0.0075 in.
2. The final dimension of the taper should be determined by making the nominal diameter about one-fourth of taper length away from the large end. For example, the $\frac{3}{4}$ inch mandrel should be 0.750 in at $1\frac{3}{16}$ in.
3. When you are near the final size, you should have an acceptable finish. If you do not, check with your instructor.
4. Clean up your finished mandrel.
5. Save the mandrel to turn in with the Worksheet 3 project.

WORKSHEET 3. Grinding a Bore

Materials

Any part or project requiring an internal ground surface.
Note: Your instructor may or may not wish you to do this exercise. This exercise can also be done on a tool and cutter grinder that has a cylindrical grinding attachment.

Procedure

1. Whether you use the high speed attachment on the cylindrical grinder or the cylindrical grinding attachment on a tool and cutter grinder, the method is the same.
2. The part to be internally ground is usually mounted in a precision chuck and set up to run true. The rotational speed will need to be adjusted for optimum finish and so that the work surface and wheel surface are moving in opposite directions.
3. A small diameter mounted wheel is usually used for small bore internal grinding. The shank of the wheel is held in a collet in the spindle. The wheel speed will need to be set according to diameter and within safe wheel speed. Table movement, wheel dressing, and use of coolant is similar to external grinding.
4. Use light cuts to produce good finishes. Grind to diameter.
5. Clean up the workpiece and the equipment you used.
6. Turn in the grading sheet, the mandrel, and the completed internal grinding project to your instructor for evaluation.

MACHINE TOOL PRACTICES

Name _____ Date _____

PROJECT 19. TOOL AND CUTTER GRINDING

Project Evaluation (To be filled out by the instructor):

	Grade	
	Letter	Percent
1. Follows drawing (dimensions and tolerances)	_____	_____
2. Machining finishes	_____	_____
3. Mechanism or tool operates satisfactorily	_____	_____
4. General workmanship	_____	_____
Total grade	_____	_____

Comments:

Signed: _____
(Instructor)

MACHINE TOOL PRACTICES   173

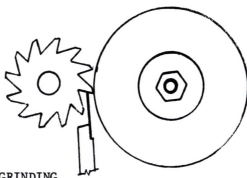

PROJECT 19. TOOL AND CUTTER GRINDING

Objectives

1. Learn to sharpen plain and slab milling cutters, and also side cutting types.
2. Learn to sharpen end mills on the end and periphery.

Outline for Study

Prior to starting each procedure for this project, study:

1. Sharpening milling cutters: Section N, Unit 10.

Procedures

Begin the procedures for this project by completing the following worksheets.

Worksheet 1. Grinding a Plain Helical Milling Cutter
Worksheet 2. Sharpening a Side Cutting cutter
Worksheet 3. Sharpening an End Mill

WORKSHEET 1. Grinding a Plain Helical Milling Cutter

Materials

Universal tool and cutter grinder and attachments
One or more dulled helical milling cutters

Procedure

1. Mount the cutter on a grinding arbor as shown in your textbook on page 699, Figures 19, 20, and 21.
2. Set up for grinding the primary clearance as explained on pages 698 to 700 in your textbook.
3. Do not attempt to grind below severe gouges and nicks in the cutter teeth. It is a more economical procedure to just remove dullness and damage that is not more than 0.005 to 0.015 in. deep. If there is a question on how much to remove, ask your instructor.
4. Show your sharpened cutter to your instructor.

WORKSHEET 2. Sharpening a Side Milling Cutter

Material

One or more side milling cutters, such as staggered tooth types

Procedure

1. If you are sharpening a staggered tooth cutter, set up a toothrest like that shown in Figure 13, page 697, of your textbook. If the teeth on the cutter are widely spaced, use a toothrest like the one on the upper right in Figure 13; but if the teeth are close together, use one like the one below it, that is, an offset toothrest.
2. The wheel should be a narrow straight type, vee-shaped to a point at the center.
3. The vee point of the toothrest should be aligned with the vee point of the wheel and centered for height with the center gage.
4. The wheelhead drop is calculated the same way as for the plain helical mill.
5. Grind the primary clearance, first in one direction on one tooth, and then back in the opposite direction with the other tooth on the other side of the toothrest.
6. Set the drop and grind the secondary clearance, if needed. Note: Some staggered tooth cutters are made so a secondary grind is not necessary.
7. The side cutting edges are not usually ground unless they have been dulled from severe use. (Grinding the side teeth of a cutter reduces its usefulness somewhat, since a nominal, or even number, width is useful for cutting grooves of a standard width, such as keyseats.) This can be easily done, using a flaring cup wheel on a tilting wheelhead (page 701, Figure 27 in your textbook). If your machine is not so equipped, an attachment called a universal workhead, such as that shown on pages 701 and 703, Figures 27 and 32, with a stub arbor clamped in a collet can be used. Set up the workhead to about 5 to 10 degrees clearance. If that attachment is used, a flicker type toothrest is used to position the teeth. If a cutter is narrowed by sharpening on the side, the marking on the side of the cutter that identifies its nominal width should be ground off and the new dimension put in its place with a tool marker.
8. Show your sharpened cutter to your instructor.

WORKSHEET 3. Sharpening an End Mill

Material

One or several dulled end mills

Procedure

1. Perhaps 80 to 90 percent of all end mill sharpening is done on the end since that is where most of the cutting is done. Breakage is usually on the outer corners. These corners can be removed by simply cutting off the end. See Figure 32, page 703 in the textbook. A damaged end may also be removed by grinding it off on a pedestal grinder, but care must be taken not to overheat the tool.
2. When a two-flute end mill has been cut off, it looks something like a thick letter "S" on the end. At this point, it must be gashed to restore the cutting edge.

3. Commercial tool sharpeners usually do this operation offhand as shown in Figure 35, page 698, to save time. It is a somewhat hazardous method, however, for the inexperienced tool grinder. Some school instructors prefer the students do this operation with an universal toolhead.
4. Whichever method you use, the point is to make grinds or gashes on each tooth so that they meet at the center of the cutter, leaving a slightly undercut portion at the center. You must avoid touching the outer cutting edges when you do this.
5. Set up the cutter in the universal workhead with a flicker toothrest on the notched plate to locate the tooth positions. Set the opposite cutting edges at the same height.
6. Use a flaring cup wheel, dressed to a sharp edge. Set the universal toolhead to the primary clearance angle of about 4 degrees and adjust the stop so the edge of the wheel comes to the center of the cutter.
7. Grind the primary angle on all teeth, just removing the undercut at the center. This will bring the two cutting edges to a single point at the center of the cutter.
8. Set the toolhead to the secondary angle of about 15 degrees.
9. Grind the secondary angles.
10. If the periphery of the end mill is in need of sharpening, use the procedures outlined in your textbook on pages 701 and 703. Ask your instructor about this procedure.
11. Turn in the grading sheet and the cutters to your instructor for evaluation.

MACHINE TOOL PRACTICES

Name _____ Date _____

PROJECT 20. FINAL TERM PROJECT

Project Evaluation (To be filled out by the instructor):

	Grade	
	Letter	Percent
1. Follows drawing (dimensions and tolerances)	_____	_____
2. Machining finishes	_____	_____
3. Mechanism or tool operates satisfactorily	_____	_____
4. General workmanship	_____	_____
Total grade	_____	_____

Comments:

Signed: _____
(Instructor)

PROJECT 20. FINAL TERM PROJECT

Objectives

1. Learn to apply the training received thus far in such areas as machine technology, mathematics, and drafting to the design and manufacture of a simple mechanical device.
2. Learn how to plan the design of a mechanism.
3. Be able to build a complex mechanism from standard stock material.

Procedure

By now you should have completed all of the units of study in your textbook; no further study is necessary in the text except for referral to refresh your memory. However, further quest for information should never come to an end in your career. While you plan your final project, you may need to consult many reference books to find needed information   It would be helpful if a drafting course could be taken at a convenient time prior to the term in which you plan to build your project. The project design could then be an advanced drafting project as well, and you would have the help of your drafting instructor.

The project should be planned with the permission and help of your machine shop instructor, who will be able to tell you if you are able to complete it or if it is feasible to try to make at all. You should start out with an idea. Make some rough sketches of the idea and show them to your instructor. You may need to modify or completely change it many times until it is workable. When all concerned accept your sketches, make an assembly drawing to scale and continue the process of modification. When it is satisfactory, then make the detail drawings. Only then will you be ready to begin making the parts.

Your instructor may have certain criteria for the final project. Here are some.

1. It must be a mechanism that requires the use of most, if not all, machine tools in the shop.
2. It must have many machine elements such as shafts, threads, gears, and bearings or bushings
3. It must reflect the ability of the particular student.

Some ideas for final projects are:

1. Reciprocating power hacksaw
2. Cutoff bandsaw
3. Wood lathe
4. Drill press
5. Arbor press
6. Geared winch
7. Dovetail machine slides

MACHINE TOOL PRACTICES

Name _____ Date _____

PROJECT 3. PIN PUNCH SET (Alternate to Center Punch)

Project Evaluation (To be filled out by the instructor):

	Grade	
	Letter	Percent
1. Follows drawing (dimensions and tolerances)	_____	_____
2. Machining finishes	_____	_____
3. Mechanism or tool operates satisfactorily	_____	_____
4. General workmanship	_____	_____
Total grade	_____	_____

Comments:

Signed: _____
(Instructor)

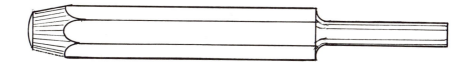

PROJECT 3. PIN PUNCH SET (Alternate to Center Punch)

Objectives

1. Learn to identify and use tool steels.
2. Learn to turn a taper using the compound rest.
3. Be able to harden and temper a tool steel correctly.

Outline for Study

Prior to starting each procedure for this project, study and complete Post-Tests for:

1. Identifying metals: Section D, Units 1 and 2.
2. Machining the punches: Section I, Unit 13.
3. Hardening and tempering the punches: Section D, Unit 3.

Procedures

Worksheet 1. Machining the Punches
Worksheet 2. Hardening and Tempering the Punches

WORKSHEET 1. Machining the Punches

Materials

15 in. of $\frac{3}{8}$ in. dia. hexagonal or octagonal S-5 tool steel or SAE 1095 carbon steel

11 in. of $\frac{1}{2}$ in. dia. hexagonal or octagonal S-5 tool steel or SAE 1095 carbon steel

Procedure

Begin by selecting the material from the steel rack. It can be $\frac{3}{8}$ in. and $\frac{1}{2}$ in. octagon or hexagon stock as measured across the flats. Be sure it is tool steel and not mild steel; if you are not sure, as your instructor or spark test a sample if you are familiar with spark testing procedures. This tool steel may be S-5 shock resistant tool steel or SAE 1095 which contains sufficient carbon to harden by quenching. In the as-rolled condition, this tool steel may have hard spots and so will quickly dull a band saw. For this reason, you should cut off this material with a hand hacksaw. Cut off all eight pieces $\frac{1}{16}$ in. over the given length on the drawing.

1. Set up a four-jaw chuck if the stock is 8-sided, or a three-jaw chuck if it is 6-sided.

2. Extend the piece out from the chuck the length of the part to be machined, plus clearance for the tool holder. Set up the stock to run true in the chuck.
3. Take fairly light cuts with a right-hand roughing tool since the work overhang prohibits deep cuts. Special care must be taken on the end of the No. 1 punch. It finishes to 0.061 in., so if you are taking even very light cuts, the last cut often breaks the end off. A method that works better is to first reduce the diameter to about $\frac{1}{4}$ in. with a fairly sharp-nosed roughing tool. Then take several cuts for a short distance (about $\frac{1}{4}$ in.) along the punch end, leaving the diameter about 0.001 in. over the finish size. Then proceed along the next $\frac{1}{4}$ in., reducing the remaining length of the punch end in the same manner. File <u>very</u> lightly and carefully to size, rotating the part at a high RPM. If you bend it while filing, it will break off. Be sure to leave enough material for the radius.
4. Grind a $\frac{1}{8}$ in. radius tool. Check it with a radius gage. This tool should have zero rake and a standard relief angle. It should be set up on center. A fillet radius is one of the most important cutting operations that a machinist must be able to do correctly. It should not be an undercut as this sould form a groove that could weaken the part and cause it to break in service. It should be smooth with no tool marks which can cause cracks to develop in this sensitive location.
5. Using a lower speed than for single point turning and cutting oil, cut the radius. Machine the shoulder to length (if you are working on the No. 4 through No. 8 punches).
6. Set the compound to 10 degrees and rough out and finish the tapered part to length (if you are working on the No. 1 through No. 3 punches).
7. Turn the part end for end in the chuck and set up for running true.
8. Turn the 10 degree taper as shown in the drawing.
9. Free-hand turn the $\frac{3}{8}$ in. radius and check with a radius gage. Finish with a file and abrasive cloth.
10. Remove the punch from the lathe and stamp your name or initials on the shank, if desired.

WORKSHEET 2. Hardening and Tempering the Punches

1. Grasp the shank end of the punches with tongs and place them on a brick or raised block in a furnace that has been heated to 1550°F.
2. When the punches have become the same color as the furnace bricks, remove one punch at a time and quench in water or oil, depending the type of tool steel. The No. 1 punch should be quenched in oil in any case since a water quench is too severe for this small mass and could cause it to crack.
3. Remove any oil from the punches and polish off the scale on the machined surfaces with abrasive cloth until a bright metallic surface shows. Be careful to keep any oil or fingerprints off this surface.
4. Place the shank part of the punches on a hot plate and, if necessary, move them around to obtain an even coloring. The shank end where it will be struck with a hammer should become violet or blue, but the punch end should color a deep gold with a mottling of purple where the punch end joins the shank.
5. When these colors are obtained, quickly plunge the punch in water so it will not become any softer.
6. Turn in the punch set for grading. Leave the tempering colors on the punches so your instructor can evaluate the heat treatment.

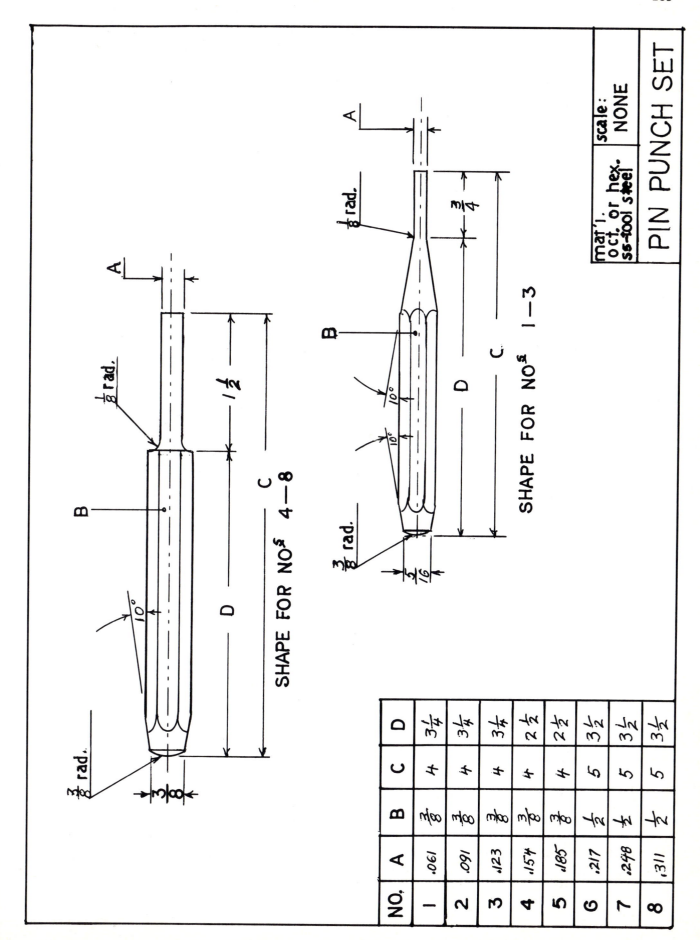

MACHINE TOOL PRACTICES

Name _____ Date _____

PROJECT 14. WHEEL PULLER (Alternate to Hydraulic Jack)

Project Evaluation (To be filled out by the instructor)

	Grade	
	Letter	Percent
1. Follows drawing (dimensions and tolerances)	_____	_____
2. Machining finishes	_____	_____
3. Mechanism or tool operates satisfactorily	_____	_____
4. General workmanship	_____	_____
Total grade	_____	_____

Comments:

Signed: _____

(Instructor)

MACHINE TOOL PRACTICES 189

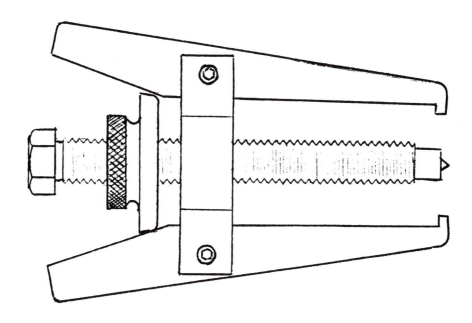

PROJECT 14. WHEEL PULLER (Alternate to Hydraulic Jack Project)

Objectives

1. Be able to turn a disc shape on a lathe and tap an internal thread.
2. Learn to mill slots on a horizontal milling machine.
3. Be able to use a vertical bandsaw.
4. Use a steady rest on a lathe.
5. Case harden mild steel.

Outline for Study

Prior to starting each procedure for this project, study and complete Post-Tests for:

1. Making the body: Section K, Units 1 through 7; Section J, Units 1 through 4; Refer to Section I, Unit 9.
2. Making the nut: Refer to Section I, Unit 9.
3. Making the screw: Refer to Section I, Units 10, 11, and 14.
4. Making the legs: Refer to Section G, Units 3 and 4; Section D, Unit 3.

Procedures

Begin the procedures for this project by completing the following worksheets.

Worksheet 1. Making the Body
Worksheet 2. Making the Nut
Worksheet 3. Making the Screw
Worksheet 4. Making the legs

WORKSHEET 1. Making the Body

Materials

Three $\frac{1}{4}$-20 Soc. HD capscrews $\frac{7}{8}$ in. of $3\frac{1}{2}$ in. HRMS round bar

Procedure

1. Saw the material for the body from $3\frac{1}{2}$ in. MS round stock. Cut off a piece about $\frac{7}{8}$ in. long.
2. Set up a three-jaw chuck with the jaws set to grip the outside of this piece. Place the material in the chuck with the jaws gripping about $\frac{1}{8}$ in. True up the piece by measuring from the chuck face with a rule in several locations.
3. Take a facing cut to clean up one side.
4. Turn the OD to the size given in the drawing. Machine as close to the chuck jaws as possible, staying clear of the jaws with the tool. Chamfer the edge.
5. Remove the part from the chuck and apply layout dye to the finished surface.
6. To lay out the part, find the center and scribe a centerline a across the surface (Figure 1). Prick punch the center.
7. Set the dividers to approximately $1\frac{1}{2}$ in. Put one leg in the prick punch mark and scribe an arc of about 180 degrees.
8. With the dividers at the same setting, place one leg where the centerline and the arc coincide. This will mark points 60 degrees on either side of the centerline (d and e, Figure 2).

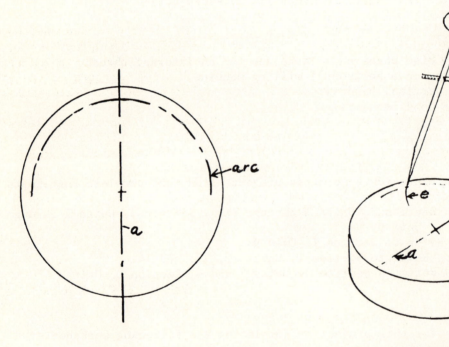

Figure 1.　　　　　　　　　　Figure 2.

9. Scribe a centerline (b and c, Figure 3) through the center to the two short arcs (d and e).
10. On the four lines that now extend to the outside edge of the circle, measure in $\frac{7}{8}$ in. and scribe a short line perpendicular to the centerline to outline the bottom of the slots.
11. Measure $\frac{1}{8}$ in. on either side of the centerlines and scribe parallel lines to meet the short cross line (Figure 3).

12. Scribe parallel lines $\frac{1}{4}$ in. from either side of the slots. Mark their length from the outer edge $\frac{11}{16}$ in.
13. Connect the ends of these lines to form the remainder of the outline of this part. The layout is now complete.
14. Replace the part in the lathe chuck with the finished face in. Grasp it this time as far back into the jaws as it will go. Be sure the seating surfaces of the jaws are clean and free from chips.
15. Face to the thickness given in the drawing. This should remove the $\frac{1}{8}$ in. section where the jaws were holding the part in the last operation. Chamfer the OD.
16. Centerdrill and then drill through with a pilot drill or a drill slightly smaller than the tap drill.
17. Drill through with a tap drill for $\frac{3}{4}$-16 NF thread.
18. Chamfer the hole.
19. Set up the tap with the dead center supporting it and the tap handle resting on the carriage or on the compound, not on the ways. Release the spindle so it is free wheeling. Turn the chuck by hand to make the thread while following with the dead center by gently turning the tailstock handwheel. <u>Do not turn on the machine for this operation.</u> Use sulfurized cutting oil. Remove the work when finished and clean the machine.
20. To continue with this project, you should be familiar with the use of a small horizontal milling machine. You will need to be able to set up the cutter, vise, and work. You must also set your speeds and feeds. You will need to study Section J, Units 1 and 2, to be ready for this operation.
21. Cut a slot first, then make the two cuts on each side of the slot (Figure 4). Use cutting fluid. Some machines are capable of doing this in one cut; however, some light duty machines may only be capable of .200 to .400 in. depth per pass.

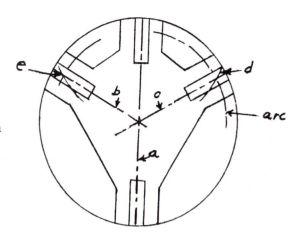

Figure 3.

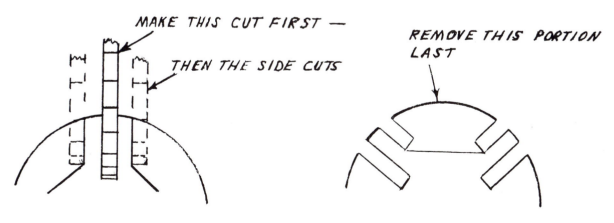

Figure 4.

Figure 5.

22. Turn the work and set up the next centerline with a square head. Repeat the cutting procedure in step 21.
23. When all four slots and side cuts have been made, rotate the work so that the line on the spaces in between is level or parallel with the solid jaw of the vise.
24. Remove all excess material by taking cuts with the same milling cutter (Figure 5). Note: An alternate method is to saw out the excess material on a vertical bandsaw.
25. When the milling is finished, remove the part, deburr and chamfer all edges.
26. Lay out for the four holes in the ears for the capscrews and center punch them.
27. Set up in the drill press with one ear level with the table.
28. Drill through both halves of the ear with a $\frac{1}{4}$-20 tap drill.
29. Drill through the <u>top ear only</u> with a $\frac{1}{4}$ in. drill.
30. Rotate the piece to the next ear and repeat steps 27 through 29.
31. After all four ears are drilled, deburr and tap through the ears with a $\frac{1}{4}$-20 tap. Clean and apply a light film of oil.

WORKSHEET 2. Making the Nut

Material

$1\frac{1}{4}$ inches of $2\frac{1}{2}$ HRMS round

Procedure

1. Cut off a piece of mild steel round stock slightly larger than $2\frac{1}{4}$ in. diameter. The next size larger may be $2\frac{1}{2}$ in. HRMS round bar. The length should be $1\frac{1}{4}$ in.
2. Since there are some very heavy tool pressures in this operation such as knurling, grooving, and drilling, it is best to set up the work in a four-jaw chuck. Leave $\frac{7}{8}$ in. extending out of the chuck and roughly true up the work with no more than 0.010 in. runout.
3. The operation with the greatest work force should be done first so that if the work moves it can be more easily corrected. Therefore, the knurling should be done first, but the part must first be machined to the knurl diameter.
4. When you have turned the nut to the knurl diameter in as far as the shoulder ($\frac{9}{16}$ in.), then set up a medium knurl and knurl the part.
5. Grind a tool bit with a $\frac{1}{8}$ in. radius.
6. Machine the groove with the lowest speed on the lathe. Use cutting oil. This plunge cut, like parting off, may produce some chatter. If the feed is too light, it will almost always chatter, so feed in by hand at a rate that will just produce a chip. Too much feed will also cause chatter or jamming of the tool in the work, causing the work to come out of the chuck or the machine to stop.
7. Face the piece and chamfer both sides of the knurled part. Finish turn both faces and the shoulder again to $\frac{9}{16}$ in. from the end to correct any runout.
8. Centerdrill, pilot drill, and tap drill with the same procedure as with the body.
9. Tap $\frac{3}{4}$-16.
10. Remove the nut from the chuck and change to a three-jaw chuck.
11. Chuck a threaded stub mandrel having a $\frac{3}{4}$-16 NF thread.
12. Screw on the nut with the knurled part next to the chuck.

13. Turn the OD to $2\frac{1}{4}$ in. Face to the $\frac{13}{16}$ in. length. Mark a 2 in. diameter line on the face.
14. Free hand machine a slight radius from the OD to the 2 in. line as shown on the drawing.
15. Finish the radius with a file and deburr the edge nearest the chuck. Chamfer the threaded hole. Clean and apply a light film of oil.

WORKSHEET 3. Making the Screw

Material

$7\frac{1}{2}$ inches of 1 in. hexagonal stock

Procedure

1. Cut off a piece of 1 in. heagonal stock $7\frac{1}{2}$ in. long. The extra length is for turning off the center.
2. Chuck the piece in a three-jaw chuck and face both ends. Chamfer one end to a point slightly below the flats on the hexagon stock. Centerdrill the other end. Chuck $\frac{1}{2}$ in. of the chamfered end and support the other end in the tailstock center.
3. Turn the diameter to size. Check for tapering and correct if needed.
4. Set up for threading and cut the $\frac{3}{4}$-16 thread. Check for fit with the nut you have just made. Turn the $\frac{1}{2}$ in. diameter. This should now be about $\frac{3}{8}$ in. longer than it will be when finished.
5. With one end still in the chuck, set up a steady rest on the other end at the $\frac{1}{2}$ in. diameter. Move the tailstock away. Be sure the carriage is to the right of the steady rest so you can turn the end of the piece.
6. Machine off the center hole and face to length, being careful to leave material in the center for the point.
7. Remove the part from the lathe and case harden by the roll method explained in Section D, Unit 3.
8. Clean up the part and apply light oil.

WORKSHEET 4. Making the Legs

Material

$18\frac{1}{2}$ inches of $\frac{1}{4}$ x 1 in. HRMS

Procedure

1. Cut off three pieces of $\frac{1}{4}$ x 1 in. MS $6\frac{1}{16}$ in. long.
2. Lay out as shown on the drawing.
3. Drill the $\frac{1}{4}$ in. holes. Chamfer.
4. Saw along the layout lines with the vertical band saw, leaving a minimum of material to file. Use a push stick for safety when sawing.
5. Finish file the legs and chamfer all edges.
6. Case harden all three legs by the roll method. Clean and apply a light film of oil.

Procedure for Assembly

1. Obtain three $\frac{1}{4}$-20 socket head capscrews $\frac{3}{4}$ in. long. Assemble the legs on the body.
2. Install the nut on the screw and turn the screw into the body.
3. This tool may also be used in the alternate position with two legs opposed.
4. Turn in your grading sheet and completed wheel puller for your instructor's evaluation.

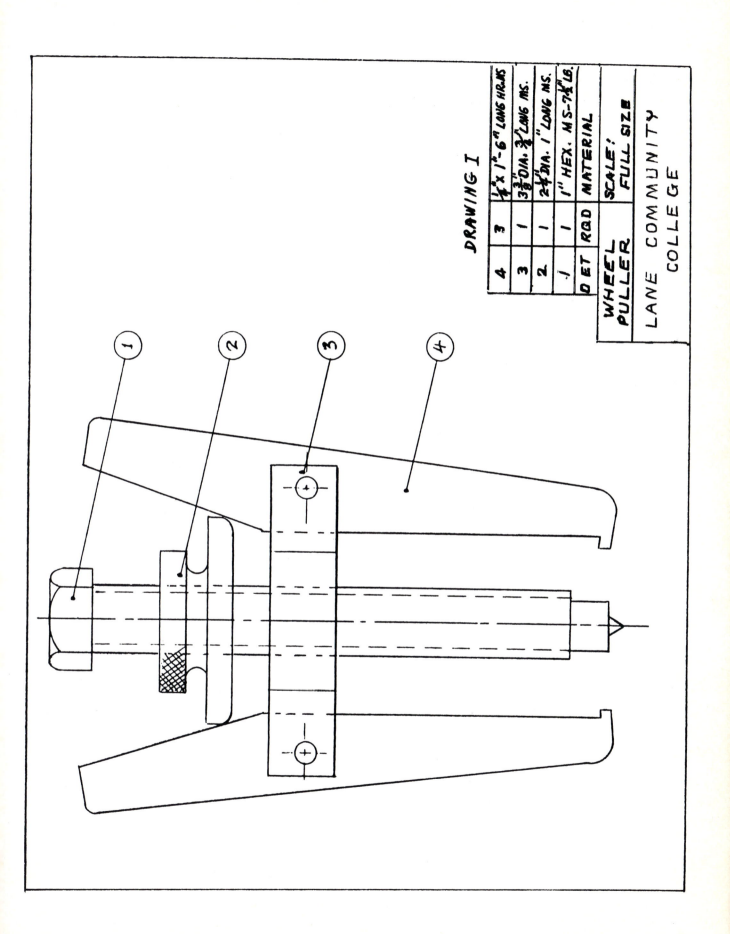

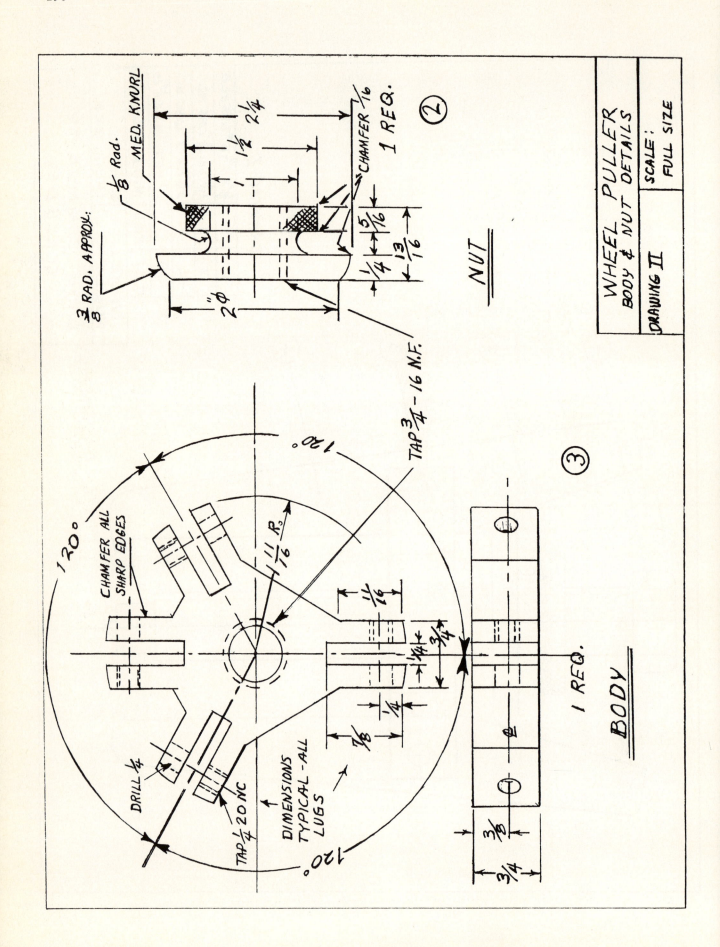

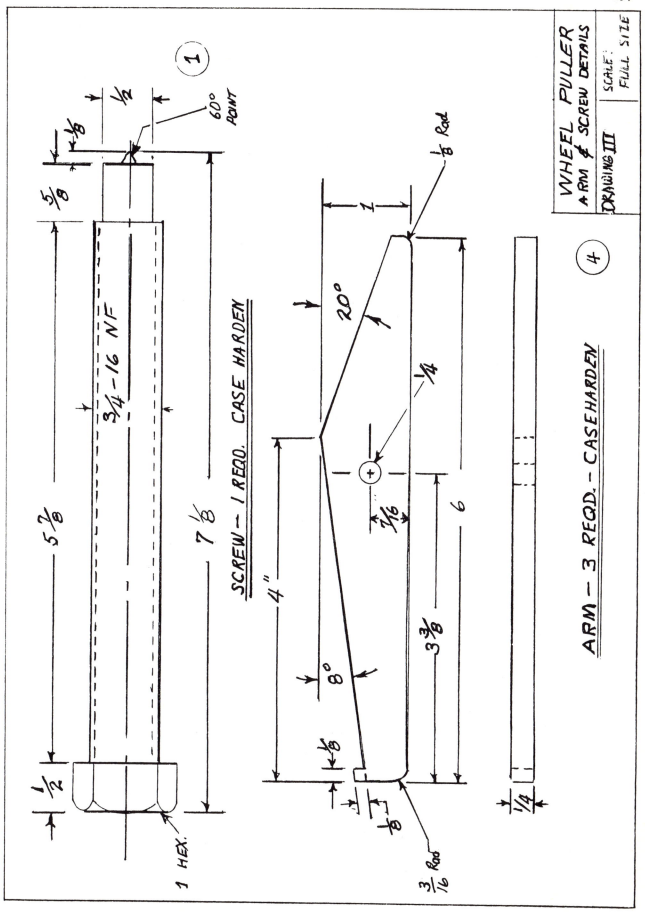

MACHINE TOOL PRACTICES

Name _____ Date _____

PROJECT 16. MACHINE VISE   (Alternate to Precision Vise Project)

Project Evaluation (To be filled out by the instructor)

	Grade	
	Letter	Percent
1. Follows drawing (dimensions and tolerances)	_____	_____
2. Machining finishes	_____	_____
3. Mechanism or tool operates satisfactorily	_____	_____
4. General workmanship	_____	_____
Total grade	_____	_____

Comments:

Signed: _____
(Instructor)

MACHINE TOOL PRACTICES   201

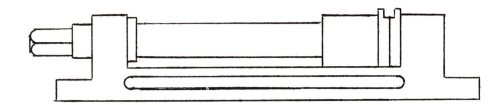

PROJECT 16. MACHINE VISE   (Alternate)

Objectives

  1. Learn to square a block and to machine steps on a horizontal mill.
  2. Be able to drill and ream holes and machine slots on a vertical mill.
  3. Surface grind the jaw inserts.
  4. Be able to make an Acme screw and nut.

Outline for Study

  Prior to starting each procedure for this project, study and complete Post-Tests for:

  1. Machining the vise body: Section K, Units 4, 6, and 7.
  2. Milling the slots and pocket: Section J, Units 5 and 6.
  3. Making the movable jaw: Refer to Section J.
  4. Grinding the jaw inserts: Section N, Unit 6.
  5. Making the screw and nut: Refer to Section I, Unit 16.

Procedures

  Begin the procedures for this project by completing the following worksheets.

  Worksheet 1. Machining the Vise Body
  Worksheet 2. Milling the Slots and Pocket
  Worksheet 3. Making the Movable Jaw and Inserts
  Worksheet 4. Grinding the Jaw Inserts
  Worksheet 5. Making the Screw and Nut

WORKSHEET 1. Machining the Vise Body

Materials

  $8\frac{1}{4}$ inches of 2 x 4 in. HRMS bar   8 inches of $\frac{1}{2}$ x $1\frac{1}{2}$ in. HRMS bar

Procedure

  1. Cut off the 2 by 4 inch stock and two pieces of $\frac{1}{2}$ by $1\frac{1}{2}$ inch bar 4 inches long.
     (Drawings for this project may be found at the end of these worksheets.)
  2. Set up the block in a horizontal mill and square it as explained on pages 579 to
     582 in the textbook. Leave it oversize for finishing, about 0.050 in.
  3. Square the ends, leaving the block about 0.050 in. long for finishing cuts.
  4. Make the chamfer for the welds on both ends of the block.
  5. Mill the chamfer on both hold-down straps and machine the ends square and to the
     oversize width of the block. Do not cut the slots in the straps at this time.

6. Clamp the straps and block to a flat surface and tack weld the upper side. The block should be preheated before welding on it because of its mass. Turn it over and weld in the groove. Alternate passes on each side to prevent warpage.
7. Set the block back in the milling vise for milling out the center portion. This is best done with a large staggered tooth cutter, about $\frac{3}{4}$ in. wide.
8. Two or more roughing cuts may be needed to machine to depth. Leave 0.030 in for finish. Make the finish cuts. If possible in the same setup, take a clean up cut on the top of the hold-down straps, leaving the weld intact. Also take a side cut to finish the outside end of the vise body, leaving the weld intact. Hold all dimensions within a tolerance of plus or minus 0.005 in.
9. Turn the vise body over and support the surface you have just machined with parallels.
10. Make the finish cut on the bottom, cleaning up the weld.
11. Set up and finish mill the sides.
12. The vise body may now be taken to a radial arm drill press and clamped to an angle plate or to the side of the drill table. The $\frac{7}{8}$ in hole in one end may now be drilled and reamed. Be sure to chamfer both ends of the hole.
13. Drill and counterbore the other end. The vise surface is completed. No further finishing (such as surface grinding) should be needed, except for deburring and chamfering the sharp edges. The holes for the jaw insert bolts may be drilled together with the inserts to insure alignment.

WORKSHEET 2. Milling the Slots and Pocket

Materials

Vertical milling machine $\frac{3}{4}$ in. center-cutting end mill

Procedure

1. The vise body should now be taken to a vertical milling machine and set up in a milling vise, bottom up.
2. The slots in the hold-down clamps may now be cut and also the $\frac{3}{4}$ in. slot in the vise body. The pocket should also be milled in this setting.
3. Turn the vise body on its side and mill the hold-down slots.
4. This completes the vise body.

WORKSHEET 3. Making the Movable Jaw and Inserts

Materials

4 inches of $1\frac{1}{2}$ x 2 in. HRMS bar   8 inches of $\frac{5}{16}$ x $1\frac{1}{2}$ in. spring steel flat bar or equivalent

Procedure for Movable Jaw

1. Square the bar on the horizontal mill as shown in your text on pages 579 to 582. Leave it slightly oversize.
2. Finish the bottom side and mill the steps. Make sure the $\frac{3}{4}$ inch guide fits the slot you have milled in the body. Measure the slot and make the guide 0.003 in. smaller for a sliding fit.
3. Finish milling the other sides and ends to the dimensions given on the drawing. Hold to a tolerance of $\pm$ 0.003 in.

4. The drilling, tapping, and countersinking may be done on the vertical mill or on a drill press. The $\frac{9}{16}$ in. socket for the screw should be reamed to size. If you wish, you may wait to drill the holes that hold the jaw inserts until the inserts are finished; they can be clamped together and drilled for better alignment. Drill the $\frac{1}{4}$ in. pin holes and their $\frac{1}{8}$ in. knockout holes before the $\frac{9}{16}$ in. hole is drilled.
5. The movable jaw is now finished.

Procedure for the Jaw Inserts

1. Since the material for the jaw inserts is spring steel, it tends to workharden very quickly if the cutting speed is too high, the feed is too low, or the tool is dull. The cutting speed for this material should be about 30 to 40 FPM. On many machines, this is the lowest possible RPM. Use cutting oil.
2. Cut 8 inches of bar and leave it this length until all the milling is done on the sides.
3. Set up on a horizontal mill in a vise for milling the grooves.
4. Take a clean-up cut with a slab mill; then change to a 60 degree single angle cutter or a 120 degree double angle cutter if one is available.
5. With a single angle cutter, one side of the angle is first cut, then the cutter is reversed on the arbor and the other side is cut.
6. The vise can be rotated 90 degrees and the short vee grooves can be cut without removing the work from the vise.
7. Now change to a side cutting cutter, and set up the bar on its edge. Mill the sides.
8. Now mill the step, measuring from the finished side.
9. Set up to mill the other flat side that is still unfinished and mill it to size, leaving about 0.020 in. for grinding.
10. Saw the bar into two parts and mill the ends square and to length.
11. Lay out, drill, and tap the $\frac{1}{4}$-28 holes. The drilling of the movable jaw and vise body may be done together.
12. All of the machining must be done before the next step.
13. Harden and temper the jaws as explained in Section D, Unit 3, of your textbook.

Procedure for Retainer

1. At this point the retainer plate should be made as shown on the drawing. It should slide freely in the pocket in the vise body and not extend below the bottom of the vise body.
2. The $\frac{1}{4}$ inch holes should be drilled and countersunk so that the screw heads are sunk below the surface.

WORKSHEET 4. Grinding the Jaw Inserts

Materials

Surface grinding machine and accessories

Procedure

1. Grind the flat sides of the jaw inserts as explained in Section N of the textbook.

2. Clamp the inserts into a precision vise or an angle plate and grind the edges square and to size.

WORKSHEET 5. Making the Screw and Nut

<u>Materials</u>

$8\frac{1}{8}$ inches of $\frac{3}{4}$ in. dia. CR round   1 inch of $1\frac{1}{4}$ in. round bronze stock
4 inches of $\frac{3}{4}$ x $\frac{3}{4}$ in. keystock (for optional lever)

<u>Procedure</u>

1. Centerdrill both ends of the $\frac{3}{4}$ in. round bar.
2. Set up in a dividing head for milling the four flats. If the optional lever is to be made, then simply mill two flats as shown on the drawing and drill the hole.
3. Set up in a lathe with a four-jaw chuck on the square end.
4. Turn down the $\frac{9}{16}$ inch end and make the groove.
5. Make the Acme thread as explained in Section H, Unit 16.
6. Set up a short piece of bronze bar stock in the lathe and turn the OD to be a light press fit in the vise body.
7. Tap drill for a $\frac{3}{4}$-6 Acme thread.
8. Part off the nut and set it up again in a three-jaw chuck.
9. Run a set of Acme taps through the nut while it is chucked in the lathe. Do not use power, turn the chuck or tap by hand. Alternately, if there is no tap, make the thread with a single point Acme tool.

<u>Assembly</u>

1. Clean and oil the parts.
2. Press the nut into the vise body.
3. Bolt the insert jaws to their parts. Fasten the movable jaw and screw.
4. Take the grading sheet and finished vise to your instructor for evaluation.

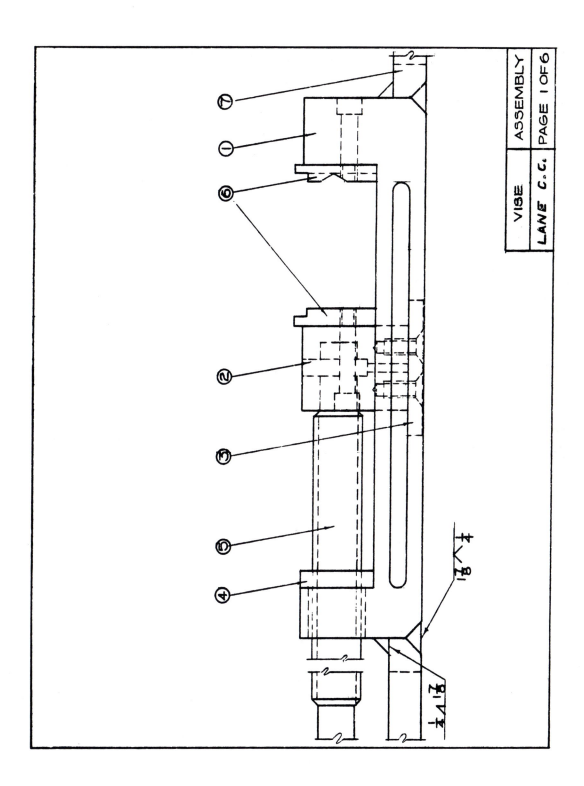

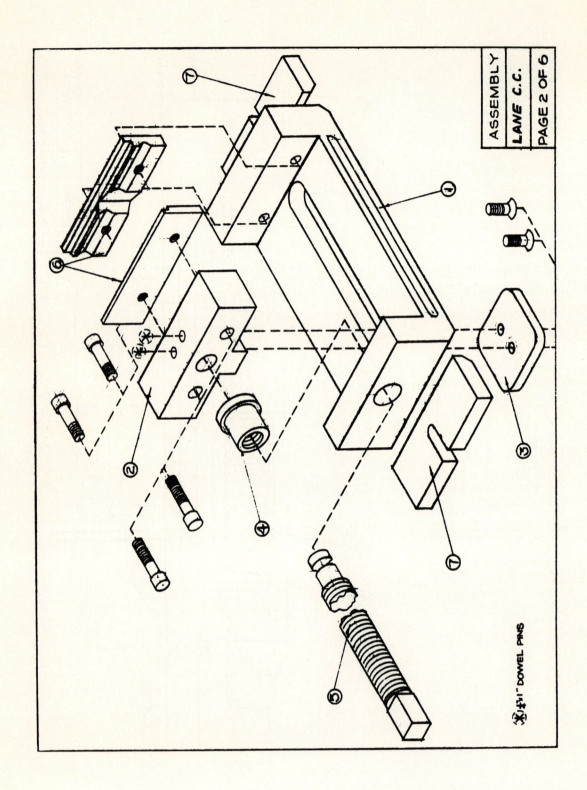

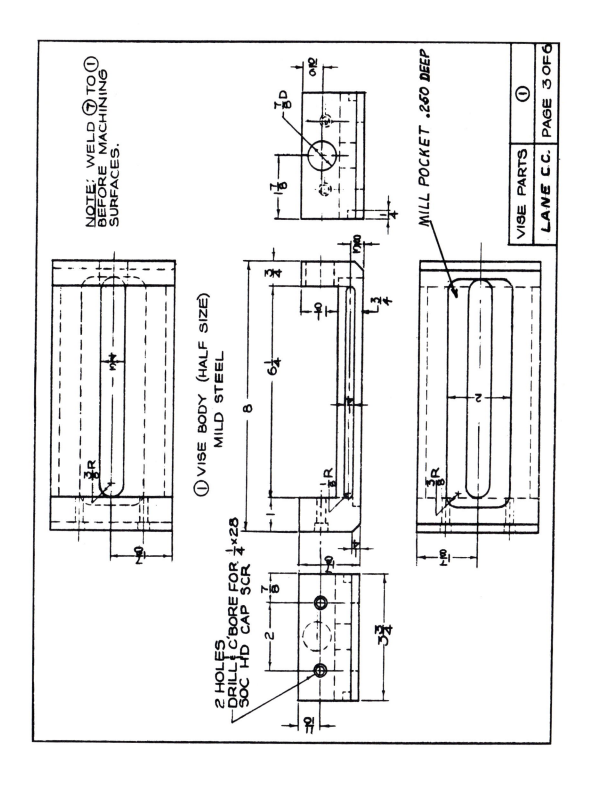

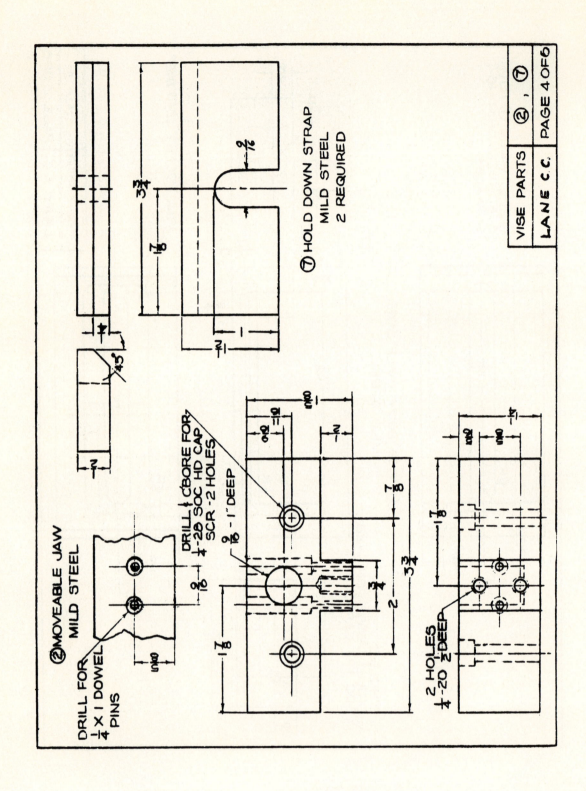

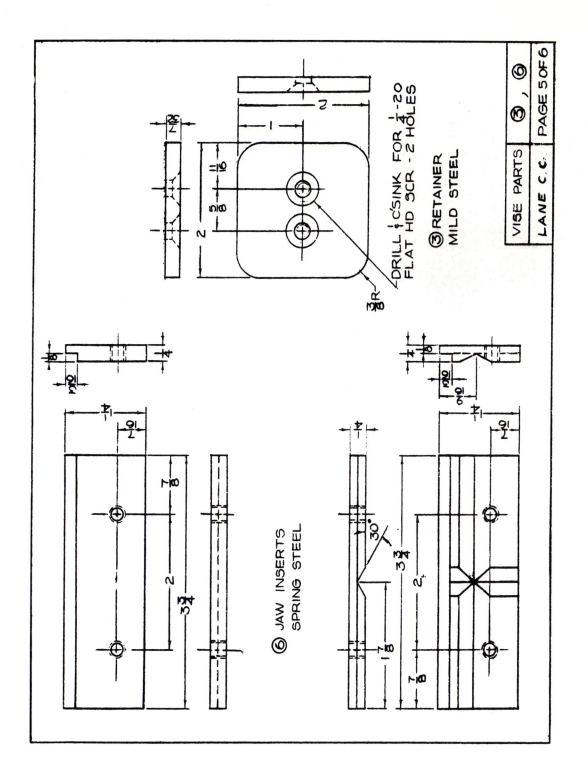

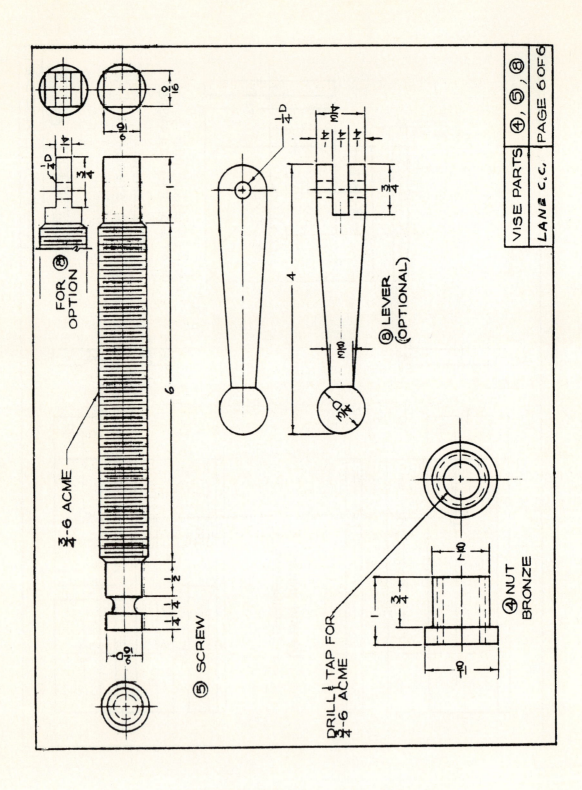

Table 1

Decimal and Metric Equivalents of Fractions of an Inch

Fractional Inch	Decimal Inch	Millimeters	Fractional Inch	Decimal Inch	Millimeters
$\frac{1}{64}$	0.015625	0.3969	$\frac{33}{64}$	0.515625	13.0969
$\frac{1}{32}$	0.03125	0.7937	$\frac{17}{32}$	0.53125	13.4937
$\frac{3}{64}$	0.046875	1.1906	$\frac{35}{64}$	0.546875	13.8906
$\frac{1}{16}$	0.0625	1.5875	$\frac{9}{16}$	0.5625	14.2875
$\frac{5}{64}$	0.078125	1.9844	$\frac{37}{64}$	0.578125	14.6844
$\frac{3}{32}$	0.09375	2.3812	$\frac{19}{32}$	0.59375	15.0812
$\frac{7}{64}$	0.109375	2.7781	$\frac{39}{64}$	0.609375	15.4781
$\frac{1}{8}$	0.125	3.1750	$\frac{5}{8}$	0.625	15.8750
$\frac{9}{64}$	0.140625	3.5719	$\frac{41}{64}$	0.640625	16.2719
$\frac{5}{32}$	0.15625	3.9687	$\frac{21}{32}$	0.65625	16.6687
$\frac{11}{64}$	0.171875	4.3656	$\frac{43}{64}$	0.671875	17.0656
$\frac{3}{16}$	0.1875	4.7625	$\frac{11}{16}$	0.6875	17.4625
$\frac{13}{64}$	0.203125	5.1594	$\frac{45}{64}$	0.703125	17.8594
$\frac{7}{32}$	0.21875	5.5562	$\frac{23}{32}$	0.71875	18.2562
$\frac{15}{64}$	0.234375	5.9531	$\frac{47}{64}$	0.734375	18.6531
$\frac{1}{4}$	0.25	6.3500	$\frac{3}{4}$	0.75	19.0500
$\frac{17}{64}$	0.265625	6.7469	$\frac{49}{64}$	0.765625	19.4469
$\frac{9}{32}$	0.28125	7.1437	$\frac{25}{32}$	0.78125	19.8437
$\frac{19}{64}$	0.296875	7.5406	$\frac{51}{64}$	0.796875	20.2406
$\frac{5}{16}$	0.3125	7.9375	$\frac{13}{16}$	0.8125	20.6375
$\frac{21}{64}$	0.328125	8.3344	$\frac{53}{64}$	0.828125	21.0344
$\frac{11}{32}$	0.34375	8.7312	$\frac{27}{32}$	0.84375	21.4312
$\frac{23}{64}$	0.359375	9.1281	$\frac{55}{64}$	0.859375	21.8281
$\frac{3}{8}$	0.375	9.5250	$\frac{7}{8}$	0.875	22.2250
$\frac{25}{64}$	0.390625	9.9219	$\frac{57}{64}$	0.890625	22.6219
$\frac{13}{32}$	0.40625	10.3187	$\frac{29}{32}$	0.90625	23.0187
$\frac{27}{64}$	0.421875	10.7156	$\frac{59}{64}$	0.921875	23.4156
$\frac{7}{16}$	0.4375	11.1125	$\frac{15}{16}$	0.9375	23.8125
$\frac{29}{64}$	0.453125	11.5094	$\frac{61}{64}$	0.953125	24.2094
$\frac{15}{32}$	0.46875	11.9062	$\frac{31}{32}$	0.96875	24.6062
$\frac{31}{64}$	0.484375	12.3031	$\frac{63}{64}$	0.984375	25.0031
$\frac{1}{2}$	0.50	12.7000	1	1.000000	25.4000

Table 2

Tap Drill Sizes for Unified and American Standard Series Screw Threads

Thread Size	Threads Per Inch	Series	Tap Drill Diameter Size	Tap Drill Diameter Inches	Percent of Full Thread	Thread Size	Threads Per Inch	Series	Tap Drill Diameter Size	Tap Drill Diameter Inches	Percent of Full Thread
0	80	NF	1.25 mm	.0492	66	6	40	NF	31	.1200	55
			1.2 mm	.0472	79				32	.1160	68
			3/64"	.0469	81				33	.1130	77
1	64	NC	1/16"	.0625	51				34	.1110	83
			53	.0595	66	8	32	NC	28	.1405	58
			54	.0550	88				29	.1360	69
1	72	NF	1/16"	.0625	58				3.4 mm	.1338	74
			1.55 mm	.0610	66				3.3 mm	.1299	84
			53	.0595	75	8	36	NF	27	.1440	55
2	56	NC	49	.0730	56				28	.1405	65
			50	.0700	69				29	.1360	77
			1.75 mm	.0689	74				3.4 mm	.1338	83
			51	.0670	82	10	24	NC	22	.1570	61
2	64	NF	1.9 mm	.0748	55				24	.1520	70
			49	.0730	64				25	.1495	75
			1.8 mm	.0709	74				26	.1470	79
			50	.0700	79	10	32	NF	19	.1660	59
3	48	NC	45	.0820	63				20	.1610	71
			46	.0810	66				21	.1590	76
			47	.0785	76				22	.1570	81
			48	.0760	85	12	28	NF	14	.1820	63
3	56	NF	44	.0860	56				15	.1800	66
			2.15 mm	.0846	62				16	.1770	72
			45	.0820	73				17	.1730	79
			46	.0810	77	12	28	NF	12	.1890	58
4	40	NC	42	.0935	57				14	.1820	73
			2.3 mm	.0905	66				15	.1800	77
			43	.0890	71				16	.1770	84
			44	.0860	80	12	32	NEF	3/16"	.1875	70
4	48	NF	41	.0960	59				13	.1850	76
			42	.0935	68				14	.1820	84
			2.3 mm	.0905	79	1/4"	20	UNC	5	.2055	67
			43	.0890	85			NC	6	.2040	71
5	40	NC	37	.1040	65				7	.2010	75
			38	.1015	72				8	.1990	78
			39	.0995	78	1/4"	28	UNF	2	.2210	62
			40	.0980	83			NF	7/32"	.2187	67
5	44	NF	36	.1065	62				5.5 mm	.2165	72
			37	.1040	71				3	.2130	80
			38	.1015	79	1/4"	32	NEF	5.7 mm	.2244	63
			39	.0995	86				2	.2210	71
6	32	NC	34	.1110	66				7/32"	.2187	77
			35	.1100	69				5.5 mm	.2165	82
			36	.1065	77	5/16"	18	UNC	17/64"	.2656	65
			37	.1040	83			NC	G	.2610	71

Table 2 (continued)

Thread Size	Threads Per Inch	Series	Tap Drill Diameter Size	Tap Drill Diameter Inches	Percent of Full Thread
			F	.2570	77
			6.4 mm	.2520	84
5/16"	24	UNF	J	.2770	66
		NF	I	.2720	75
			H	.2660	85
5/16"	32	NEF	7.3 mm	.2874	62
			7.2 mm	.2835	71
			9/32"	.2812	77
			J	.2770	87
3/8"	16	UNC	P	.3230	64
		NC	O	.3160	72
			5/16"	.3120	77
			7.8 mm	.3071	83
3/8"	24	UNF	R	.3390	67
		NF	8.5 mm	.3346	74
			Q	.3320	79
			21/64"	.3281	86
7/16"	14	UNC	3/8"	.3750	67
		NC	U	.3680	75
			23/64"	.3594	84
7/16"	20	UNF	X	.3970	62
		NF	25/64"	.3906	72
			W	.3860	79
1/2"	13	UNC	7/16"	.4375	62
		NC	27/64"	.4219	78
			Z	.4130	87
1/2"	20	UNF	11.75 mm	.4626	57
		NF	29/64"	.4531	72
9/16"	12	UNC	1/2"	.5000	58
		NC	31/64"	.4844	72
			15/32"	.4687	86
9/16"	18	UNF	33/64"	.5156	65
		NF	13 mm	.5118	70
			1/2"	.5000	86
5/8"	11	UNC	14 mm	.5512	63
		NC	35/64"	.5469	66
			17/32"	.5312	79
5/8"	18	UNF	37/64"	.5781	65
		NF	14.5 mm	.5709	75
			9/16"	.5625	87
3/4"	10	UNC	17 mm	.6693	62
		NC	21/32"	.6562	72
			41/64"	.6406	84
3/4"	16	UNF	45/64"	.7031	58
		NF	11/16"	.6875	77
			43/64"	.6719	96
7/8"	9	UNC	25/32"	.7812	65
		NC	49/64"	.7656	76
			3/4"	.7500	87
7/8"	14	UNF	13/16"	.8125	67
		NF	20.5 mm	.8071	73
			51/64"	.7969	84
1"	8	UNC	57/64"	.8906	67
		NC	7/8"	.8750	77
			55/64"	.8593	87
1"	12	UNF	15/16"	.9375	58
		N	59/64"	.9218	72
			29/32"	.9062	87

Some Symbols Used for American Threads Are:

Symbol	Reference
NC	American National Coarse Thread Series
NF	American National Fine Thread Series
NEF	American National Extra Fine Thread Series
NS	Special Threads of American National Form
NH	Am. Natl. Hose Coupling and Fire Hose Coupling Thread
NPT	American Standard Taper Pipe Thread
NPTF	American Standard Taper Pipe Thread (Dryseal)
NPS	American Standard Straight Pipe Thread
ACME	Acme Threads—(Acme-C) Centralizing—(Acme-G) General Purpose
STUB ACME	Stub Acme Threads
V	A 60° "V" thread with truncated crests and roots. The theoretical "V" form is usually flattened several thousandths of an inch
SB	Manufacturers Stovebolt Standard Thread

Symbols Used for Unified Threads Are:

Symbol	Reference
UNC	Unified Coarse Thread Series
UNF	Unified Fine Thread Series
UNEF	Unified Extra Fine Thread Series

Table 3

Metric-Inch Conversion Table

Milli-meters	Decimal Inches	Milli-meters	Decimal Inches	Milli-meters	Decimal Inches	Milli-meters	Decimal Inches	Milli-meters	Decimal Inches	Milli-meters	Decimal Inches
.1	.00394	11	.4331	29	1.1417	47	1.8504	65	2.5590	83	3.2677
.2	.00787	12	.4724	30	1.1811	48	1.8898	66	2.5984	84	3.3071
.3	.01181	13	.5118	31	1.2205	49	1.9291	67	2.6378	85	3.3464
.4	.01575	14	.5512	32	1.2598	50	1.9685	68	2.6772	86	3.3858
.5	.01968	15	.5905	33	1.2992	51	2.0079	69	2.7165	87	3.4252
.6	.02362	16	.6299	34	1.3386	52	2.0472	70	2.7559	88	3.4646
.7	.0275	17	.6693	35	1.3779	53	2.0866	71	2.7953	89	3.5039
.8	.0315	18	.7087	36	1.4173	54	2.1260	72	2.8346	90	3.5433
.9	.03541	19	.7480	37	1.4567	55	2.1653	73	2.8740	91	3.5827
1	.0394	20	.7874	38	1.4961	56	2.2047	74	2.9134	92	3.6220
2	.0787	21	.8268	39	1.5354	57	2.2441	75	2.9527	93	3.6614
3	.1181	22	.8661	40	1.5748	58	2.2835	76	2.9921	94	3.7008
4	.1575	23	.9055	41	1.6142	59	2.3228	77	3.0315	95	3.7401
5	.1968	24	.9449	42	1.6535	60	2.3622	78	3.0709	96	3.7795
6	.2362	25	.9842	43	1.6929	61	2.4016	79	3.1102	97	3.8189
7	.2756	26	1.0236	44	1.7323	62	2.4409	80	3.1496	98	3.8583
8	.3150	27	1.0630	45	1.7716	63	2.4803	81	3.1890	99	3.8976
9	.3543	28	1.1024	46	1.8110	64	2.5197	82	3.2283	100	3.9370
10	.3937										

Table 4

Inch-Metric Measures

Linear Measures

English	Metric
1 mile = 1760 yards = 5280 feet	10 millimeters (mm) = 1 centimeter (cm)
1 yard = 3 feet = 36 inches	10 centimeters = 1 decimeter (dm)
1 foot = 12 inches	10 decimeters = 1 meter (m)
1 inch = 1000 mils	1000 meters = 1 kilometer (km)

Conversion Factors

1 inch = 2.54 cm = 25.4 mm	1 millimeter = .03937 inch
1 foot = .3048 meter	10 millimeters (cm) = .3937 inch
1 yard = .9144 meter	1 meter = 39.37 inches
	3.2808 feet
	1.0936 yards
1 mile = 1.6093 km	1 kilometer = 1093.6 yards or .62137 mile

Table 5

Useful Formulas for Finding Areas and Dimensions of Geometric Figures

A = Area; S = Side; D = Diagonal of circumscribed circle; d = Height of circular segment (distance cut into round stock to make 3 flats); h = Height of triangle.

Triangle (equilateral)

$A = Sh/2$
$S = D \times .866$
$d = D \times .25$
$h = D - d$

A = Area; S = Side; D = Diagonal; R = Radius of circumscribed circle; r = Radius of inscribed circle; d = Height of circular segment (distance cut into round stock to make 8 flats).

Octagon

$A = 4.828 S^2$
$S = D \times .3827$
$R = 1.307S = 1.082r$
$r = 1.207S$
$d = D \times .038$

A = Area; S = Side; D = Diagonal; d = Height of circular segment (distance cut into round stock to make 4 flats).

Square

$A = S^2 = \frac{1}{2} d^2$
$S = D \times .7071$
$D = S \times 1.4142$
$d = D \times .14645$

A = Area; D = Diameter; R = Radius; C = Circumference

Circle

$A = \pi R^2 = 3.1416 R^2 = .7854 D^2$
$C = 2\pi R^2 = 6.2832 R = 3.1416 D$
$R = C \div 6.2832$

A = Area; S = Side; D = Diagonal; R = Radius of circumscribed circle; r = Radius of inscribed circle; d = Height of circular segment (distance cut into round stock to make 6 flats).

Hexagon

$A = 2.598 S^2 = 2.598 R^2 = 3.464 r^2$
$S = D \times .5 = R = 1.155 r$
$d = D \times .067$
$r = .866 S$

A = Area; l = Length of arc; C = Chord length; R = Radius; λ = angle (in degrees); d = Height of circular segment (distance cut into round stock to produce a flat of a given width).

Circular segment

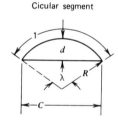

$A = \frac{1}{2}[Rl - C(R - d)]$
$C = 2\sqrt{d(2R - d)}$
$d = R - \frac{1}{2}\sqrt{4R^2 - C^2}$
$l = .01745 R\lambda$
$\lambda = \dfrac{57.296 l}{R}$

Note: To find the weight of a metal bar of any cross-sectional shape, find the area of cross section, multiply by the length of the bar in inches and by the weight in pounds per cubic inch of the material. The weights of some metals in pounds per cubic inch are as follows: steel, .284; aluminum, .0975; bronze, .317; copper, .321; lead, .409; and silver, .376.

Table 6

Allowances for Fits of Bores in Inches

Diameter (in.)	Running and Sliding Fits (free to rotate and free to slide)	Standard Fits (readily assembled)	Driving Fits (permanent assembly, light drive)	Forced Fits (permanent assembly with hydraulic press)
Up to ½	+.0005 to +.001	+.00025 to +.0005	−.0005	−.00075
½ to 1	+.001 to +.0015	+.0003 to +.001	−.0075	−.0015
1 to 2	+.0015 to +.0025	+.0004 to +.0015	−.001	−.0025
2 to 3½	+.002 to +.003	+.0005 to +.002	−.0015	−.0035
3½ to 6	+.003 to +.004	+.00075 to +.003	−.002	−.0045

Note: In this table, shaft sizes are considered as nominal and bore sizes are varied for fits, thus a negative fit is an interference (press) fit and a positive fit is a free or loose fit.

Table 7

Tapers and Corresponding Angles

Taper per Foot	Included Angle		Angle with Centerline		Taper per Inch
	Degrees	Minutes	Degrees	Minutes	
1/8	0	36	0	18	.0104
3/16	0	54	0	27	.0156
1/4	1	12	0	36	.0208
5/16	1	30	0	45	.0260
3/8	1	47	0	53	.0313
7/16	2	5	1	2	.0365
1/2	2	23	1	11	.0417
9/16	2	42	1	21	.0469
5/8	3	00	1	30	.0521
11/16	3	18	1	39	.0573
3/4	3	35	1	48	.0625
13/16	3	52	1	56	.0677
7/8	4	12	2	6	.0729
15/16	4	28	2	14	.0781
1	4	45	2	23	.0833
1 1/4	5	58	2	59	.1042
1 1/2	7	8	3	34	.1250
1 3/4	8	20	4	10	.1458
2	9	32	4	46	.1667
2 1/2	11	54	5	57	.2083
3	14	16	7	8	.2500
3 1/2	16	36	8	18	.2917
4	18	56	9	28	.3333
4 1/2	21	14	10	37	.3750
5	23	32	11	46	.4167
6	28	4	14	2	.5000

Source: Warren T. White, John E. Neely, Richard R. Kibbe, and Roland O. Meyer, *Machine Tools and Machining Practices*, John Wiley and Sons, Inc., Copyright © 1977, New York.